T H E
A U R E L I A N
O R
NATURAL HISTORY OF ENGLISH INSECTS; NAMELY,
MOTHS and BUTTERFLIES.
Together with the
P L A N T S on which they F E E D;

The Works of the Lord are Great, Sought out of all them that have Pleasure therein, Ps CXI. v. 2.

THE
AURELIAN

OR

NATURAL HISTORY OF ENGLISH INSECTS; NAMELY,

MOTHS and BUTTERFLIES.

Together with the

PLANTS on which they FEED;

A faithful Account of their respective Changes; their usual Haunts
when in the winged State; and their standard Names, as given and
established by the worthy and ingenious Society of AURELIANS.

Drawn, engraved and coloured, from the natural subjects themselves.

By *MOSES HARRIS,* 1766
Secretary to the AURELIAN SOCIETY.

Introduced by *ROBERT MAYS* 1986

Salem House Publishers
Topsfield, Massachusetts

First published in the United States
by Salem House Publishers, 1986,
462 Boston Street, Topsfield, MA 01983.

Library of Congress Catalog Card Number:
86-60176

ISBN O 88162 195 1

Published simultaneously in Great Britain by
Country Life Books, an imprint of Newnes Books,
a division of The Hamlyn Publishing Group Limited

This edition of The Aurelian was conceived,
edited, and designed by Thames Head Limited,
Avening, Tetbury, Gloucestershire, Great Britain.

Editorial and Marketing Director
Martin Marix Evans

Design and Production Director
David Playne

Designer
Nick Allen

Editor
Gill Davies

Typeset in Bembo by
Avonset, Midsomer Norton, Bath, Great Britain.

Reproduction by
Redsend Limited, Birmingham, Great Britain.

Printed by
New Interlitho, Milan, Italy

CONTENTS

The index as it appears in the Supplement to the original volume

INDEX to HARRIS's Natural History of English Moth's and Butterflies, with the Trivial Names of LINNAEUS, as far as can be collated from his works.

Insects of various Orders interspersed through this Work.

A Table of the Terms used in the Descriptions for the various Parts of the PAPILIO,
which refers to the ensuing Plates, wherein they are delineated at large.

a	HEAD	
b	Eyes	
c	Palpi	
d	Knobs of the Antennae	
e	Threads of the Antennae	
f	Tongue	
g	Thorax	
h	Shoulders	
i	Scutulum or Escutcheon	
k	Abdomen with its Anuli	
l	Tips or Apices	
m	Sector Edge	
n	Fringes	
o	Sector	
p	Abdominal Groove	
q	Tails	
r	Abdominal Corners	
s	Lowers Corners of the superior Wings	
t	Outer Corners of the inferior ditto	
u	Abdominal Edges	
w	Anus	
x	Ocelli	
y	Bar, Band or Garter	
A	Superior Wing angulated	
B	Superior Wing, smooth or even edged	
C	Inferior Wing scallaped	
D	Inferior Wing indenticulated	

The Interior Parts of the Superior Wing described.

The Parts coloured *Green*, the FAN-TENDONS, and MEMBRANES, marked in Numerical Order; viz. 1st, 2d, 3d, &c.

PAPILIONES have only five of these MEMBRANES, and six TENDONS.

The Parts coloured with *Pale Crimson*, are the TABLES.

The *Pale Blue*, shews the SECTORS.

Shoulder MEMBRANE — *Yellow*.

Slip MEMBRANE — *Pale Orange*.

Long MEMBRANE — *Pale Indian Ink*.

The Parts coloured Purple, are the SECTOR TENDONS and MEMBRANES, and it is in this Part of the Wing only, wherein one GENUS differs from another.

The Grand Tendons are coloured deep Red, and are three in Number; viz.

7 Long Tendon,
8 Principal Tendon } Grand Tendons
9 Shoulder Tendon
10 Table Tendon
11 Slip Edge
12 Bar Tendon

Interior Parts of the Inferior Wing described

Green, shews the FAN-TENDONS and MEMBRANES, which are the same in Number as in the Superior Wings.

Pale Crimson, shews the TABLE MEMBRANE.
Blue. The Sector.
Yellow. TABLE MEMBRANE or BENT, ditto.
Pale Indian Ink. LONG MEMBRANE.
Pale Orange. ACCUTE MEMBRANE.
Deep red. GRAND TENDONS, three in Number; viz.

15 POSTERIOR TENDON
16 TABLE TENDON } Grand Tendons
17 BENT or FEMORAL

18 Spur; this Part answers to that little Instrument in the Phaloena, called Spring.
13 Abdominal Tendon.
14 Long Tendon.

It is worthy to be remarked, that although the several Parts of the Inferior Wings greatly correspond with the respective Parts of the Superior, yet here are no Parts answerable to the Sector Tendons and Membranes, and which are distinguished with Purple in the Superior; but Providence seems to have supplied that Deficiency, if I may call it one, by adding another on the opposite Edge of the Wing, called THE ABDOMINAL MEMBRANE, and is coloured Purple.

FOREWORD

It gives me much pleasure to welcome this fine reproduction of the most splendid of all English entomological books, as well as to commend to the reader Robert Mays' interesting Introduction and attractively written commentaries on many of the lepidoptera so beautifully illustrated.

That Moses Harris was an acute observer and good entomological artist with a strong dedication to and love of his work, is amply testified to in *The Aurelian*. The plates in this book are particularly pleasing, with the frontispiece having particular significance, this being generally regarded as a self portrait of the author. Attired in the elegant costume of the period, he is here depicted in an enchanting sylvan scene sitting on the bank of a stream with a clap net over his knees, a chip box of butterflies and other insects in his left hand, and a pin-cushion suspended from a cord attached to his person. His right hand points to a small figure, of himself presumably, wielding a net on the opposite side of the stream.

This is a delightful book, and one from which I feel sure many will derive much fascination and enjoyment.

J. M. Chalmers-Hunt F.R.E.S.

January 1986

INTRODUCTION

There is a satisfaction when contemplating a fine example of eighteenth-century representational art which, combined with an element of romance and substantial scientific advancement, can but rarely be experienced in this modern age. In the reproduction of Harris's *magnum opus* we can admire his artistic skill, realizing at the same time that he was well in the forefront of the entomological knowledge of his era.

For a long time the exact date of Moses Harris's birth was unknown. Many reference books, in quoting the few relevant biographical facts at their disposal, indicated the year 1731. However, Dr A. A. Lisney in his *Bibliography of British Lepidoptera* (1960) related that a signed original drawing in his possession bore the inscription *Mos Harris Pinx 1783 Aged this day 53. Apl. 15,* thus finally settling the year of his birth as 1730.

As with many eighteenth-century entomologists, a cloud of obscurity surrounds the life of Moses Harris, and it must be admitted that most of our knowledge is that which he tells us himself in *The Aurelian*; his artistic and remarkably accurate observations are self-evident. Harris's original water-colour paintings for *The Aurelian* have rightly found a place in the archives of the British Museum (Natural History).

Details of his early training elude us but he must have received much encouragement for

his talents, and it seems that financial circumstances in his youth were easy, although he implies in the Preface that his education was not advanced. However, no matter how modest his claims in this respect, it is clear that he moved among educated and wealthy people, who appreciated the skill of this rising young artist. Perhaps it was his uncle of the same name, a member of the Society of Aurelians, who was his mentor from an early age, for young Moses was studying insects before the age of twelve years, an enthusiasm which, with his drawing and engraving skill and love of colour, culminated some twenty-four years later in this masterpiece.

Why did those who studied and collected butterflies and moths christen themselves 'aurelians'? It was a pleasant conceit for such an attractive name to be chosen, deriving from the Latin 'aureolus' or 'golden', so-called from the iridescent sheen upon the chrysalids of some butterfly species. The word chrysalis itself comes from the Greek word for 'gold', for the same reason. At a later date an 'aurelia' was sometimes used to describe the pupal stage of any insect.

Harris recounts in the Preface how in 1748 a disastrous fire in Change Alley, Cornhill, London, destroyed the Society's collections, regalia and books, the Aurelians just escaping with their lives. The destruction of the Swan Tavern, in which the Society rented a room from one Barton, the landlord, meant the end

of the Aurelian Society, apparently the first organized society of entomologists in England. It was some fourteen years before aurelians had sufficiently recovered from this disaster to form another Aurelian Society, this time with Moses Harris as its Secretary.

The accounts of this catastrophe included a map published by M. Payne of the White Hart in Pater-noster-Row. This was advertised on the flimsy blue covers of the only known fascicules of James Dutfield's *New and Complete Natural History of English Moths and Butterflies* (1748/9) which he also sold. It read as follows:

'Just published,
Drawn and Engrav'd on a large Scale of fifteen Yards to an Inch, and printed on a Superfine Paper.

1. A correct and regular Plan of all the Houses destroy'd and damag'd by the Fire, which begun in Exchange — Alley, Cornhill, on Friday March 25, 1748; together with the Names of all the Sufferers engrav'd on their respective Habitations.'

Included among some other contemporary publications, the *London Magazine* carried a full account of the fire, illustrated by a smaller copy of the plan which Payne had published. From this account a little more information can be gleaned about this tragic event, for apparently the outbreak occurred 'in the

Powdering Room of Mr Eldridge's Peruke Maker, near the middle of Exchange Alley', and, alas, Mr Eldridge, his wife, two children and a journeyman perished in the flames. From the map it appears the 'Swan Kitchin' actually adjoined the peruke maker's and the Tavern itself was separated from the kitchen by a very narrow alleyway.

Dr R. S. Wilkinson, the American authority on early entomologists and their methods, in referring to the account of the fire published in the *Daily Advertiser* of 26 March 1748, draws the intriguing conclusion that as the 'old-style', that is the official New Year, commenced at that time on 25 March, the Aurelians were still celebrating the New Year in the traditional way, because it was well after midnight of that day when the fire began. It was hardly likely that a regular meeting would continue to such a late hour, unless the circumstances were exceptional.

In 1749 the famous cartographer, Thomas Jefferys, whose shop was on the corner of St Martin's Lane near Charing Cross, was selling engravings by Moses Harris of a view of Halifax, Nova Scotia. Perhaps some curious researcher may discover the explanation of the apparent contradiction in Harris's own statement that he was unskilled as an engraver before working on *The Aurelian*, while it would appear that he was far from inexperienced. Unless perhaps one might suppose his uncle of the same name to be responsible, possessing similar talents, in which case he must surely have influenced his nephew in his choice of occupation.

The years following the demise of the old Aurelian Society were far from barren for Harris; although deprived of discussions to which he could bring youthful fresh ideas, he did move in entomological circles, meeting frequently with other collectors. He tells us he was 'too young and immature' to be admitted into the ranks of the earlier Society, which was his hope, at the age of twelve. We do not know the names of all those who joined that original Society, but such prominent figures in the entomological world as Joseph Dandridge, Dru Drury and Benjamin Wilkes were certainly members — Wilkes, in fact, dedicating his *Twelve New Designs of English Butterflies* (1742), to 'The Worthy Members of the Aurelian Society'.

However precocious Moses was at such an early age, it is difficult to see how he would have fitted in with men of such stature and experience. At the date of the fire Harris would have been eighteen years old, but whether the members considered he had acquired 'sufficient Sagacity' by that time is unknown. From information in Wilkes's major work *English Moths and Butterflies* (c.1749), the natural historian, David Allen, inferred that the first Aurelian Society was in existence in 1738, and his conclusions in support of this were published in 1966; it was thus shown that a sufficient number of 'ingenious and curious' entomologists gathered together long enough for the Society to function for at least ten years.

However, it is the succeeding Society of the same name with which Moses Harris was

concerned; he was appointed its first Secretary when it was formed about the year 1762, and as such he described himself in *The Aurelian*. Although it was probably due to his enthusiasm that this second Society of Aurelians was formed, after a hiatus of about fourteen years, it is recorded that it had a lifetime of only four years, thus expiring about the time of publication of *The Aurelian*. No account of a dramatic end, as with the first Society, has been traced, nor any reason for its decay and, alas, its records too have been lost.

The Aurelian was first published in 1766, but it was usual for such works to be issued periodically in numbers, such as Dutfield's book mentioned earlier, and Wilkes's *English Moths and Butterflies*, and Lisney was able to confirm that this was the intention for *The Aurelian*. He was fortunate enough to acquire the first four fascicules and, by checking the dates on the covers, was able to ascertain with a reasonable degree of certainty that the first fascicule appeared about December 1758. Wilkinson has determined the dates for the next two numbers as early the following year, and Dr H. A. Hagen in his *Bibliotheca Entomologica* (1862), drew attention to the fact that the German magazine *Göttingische gelehrte Anzeiger* for 1765 indicated that by that date fourteen plates had appeared. Very rare copies of the first four parts were also seen by Wilkinson, who was able to add a few further facts regarding the dates of their publication, which he deduced from careful examination of the manuscript insertions on the covers, making some minor corrections to Lisney.

These first four numbers of the separately issued parts were in fact incorporated in *The Aurelian* when it was first published, without of course the covers, and these, so Lisney tells us, can be identified by their shorter leaves. It is surprising that, if Hagen is correct, only four parts have ever been recorded, the printer having no subsequent parts available for similar inclusion in *The Aurelian* when it was published as a complete volume.

The original intention was that each part, with a coloured plate, should be issued monthly, but the fact that it took several years longer to finish caused a great deal of worry to Harris, particularly as the cause of the delay was not primarily his own fault; this was his first major work, and it was necessary for him to show the world what he could achieve. In defence of this unconscionable delay, he hinted that he had been badly treated by 'the unsteady and fallacious Behaviour of a Person, too nearly connected in my Concerns', making this clear in a paragraph to that effect in the Dedication. Problems such as this are not unknown to the editors and authors of scientific part works in the present day!

In the same way that pre-publication informative leaflets are now distributed by publishers, it was the practice in Harris's time to print *Proposals*, outlining details of the work and setting out the terms upon which it was to be published. Ephemera such as this are naturally extremely rare and Lisney was fortunate in having a copy pasted on the recto of the upper blank cover of the first fascicule of *The Aurelian*, which was published, as previously mentioned, in 1758; the single

sheet of the *Proposals*, which was reproduced in *A Bibliography of British Lepidoptera*, is decorated by a fine engraving by Harris of several species of lepidoptera in the perfect state and in various stages of development, as well as plants and foliage.

The text of this important announcement reads as follows:
PROPOSALS
For Engraving by SUBSCRIPTION
A Collection of PRINTS of BUTTERFLIES & MOTHS

Drawn from the Life, by M. Harris.

CONDITIONS

I   Each Plate shall Contain not only the Insect, in all its different Appearances & changes, but also the Plant on which it feeds, Colour'd in the highest Taste from Nature.

II The Proprietors propose to Publish one of the above Plates once a Month and with it a Printed Description of the Insects, & a faithful Account of their Respective Changes, their Usual Haunts when in the Fly State, & their proper Names, as given by the Worthy & Ingenious Society of Aurelians, so as to make it more easy for those who shall hereafter Attempt to Breed them.

III That each Person who shall Honour the Proprietors with their Names, shall Pay 2s. 6d. for each Number so delivered, & the Proprietors beg leave to assure their Generous Benefactors, that Nothing as to Expence, Labour, or Study shall be wanting on their sides to render the above Work worthy their Patronage.

Subscriptions are taken in by the Proprietors John Gretton Bookseller in Old Bond Street, and M. Harris at Mr Biddle's Watch Maker in New Bond Street and by all the Booksellers in London, & other parts of the Kingdom.

N.B. There will be a French Translation given with every No.

Lisney recorded that Harris described himself in the title on the blue paper covers of Numbers 2 and 3 of *The Aurelian* as a 'Painter, who has made this Part of Natural History his Study and has bred most of the Flies and Insects for these twenty years'. Young Moses claimed therefore already several years practical entomological experience before he unsuccessfully applied to join the Aurelian Society at the age of twelve. A rather longer list of booksellers than that printed in the *Proposals* was given, and in it Moses Harris gives a slightly different address for receiving subscriptions — this time it was the 'Golden Head in New Bond-Street, two Doors from Conduit-Street'; possibly it was the same building described in a way more easily identified by a stranger.

And so eventually, after nearly a decade in which there was much frustration and delay, the first issue of the first edition appeared, dedicated to members of the Society of Aurelians. It is very similar in most respects to

Self-portrait of Moses Harris (1730-c.1788) at the age of 49

The advertisement at the end of this useful guide referred to *The Aurelian* second issue as 'A new and compleat Edition with great additions, with a Table of Terms used in the descriptions, and a compleat Index, with the trivial names of Linnaeus.' The price was given as five guineas, and a note at the end indicated that 'The four additional Plates, with the Table of Terms, Index, and Trivial Names of Linnaeus, may be had separate, to complete Sets, Price 10s. 6d.'

At this point a small but important treatise by Harris must be mentioned. The full title reads as follows:
'*An Essay precedeing a Supplement to the Aurelian: Wherein are considered The Tendons and Membranes of the Wings of Butterflies; First, As useful in describing the Situation of their Spots or Markings; Secondly, Of great Assistance in discovering their different Genera. Whereunto is added, The Discovery of a particular Part, peculiar to the Moth Tribe. The whole illustrated with Copper-Plates. By Moses Harris, Miniature — Painter. London: Printed for the Author.*' A French translation of this long title also followed.

There are half a dozen leaves of text in this very rare essay, accompanied by some eight coloured engraved plates, of considerable scientific value to entomologists, as they draw attention to the neuration of wings, hitherto largely ignored, but so important when studying the classification of lepidoptera. The tendons and membranes in the plates are numbered so that text references to them could easily be made.

It has already been mentioned that *The Aurelian* Supplement was published about the year 1775, and it was clearly intended that the *Essay* should have been issued before that date. In that case surely it would have been bound in the second issue, before us now. However, Lisney, when considering this puzzling question, concluded that it was not published until 1780, and that ideal copies of a later book to be published by Harris, *An Exposition of English Insects* should, in the first issue of the first edition, 1776 (1780), have the *Essay* bound with it.

Before considering the significance of some aspects of Moses Harris's contributions in *The Aurelian* to the science of entomology, and before touching on his artistic achievements elsewhere, we shall comment briefly on the later editions of *The Aurelian* — for the success of the first edition encouraged him to proceed with a second which was published in 1778. The frontispiece and plates were the same as in the volume before us, but this time the text was also printed in French, as well as English, in double columns; a format which was continued in the two further issues of this edition, in 1778 (the same year as the first issue), and lastly in an undated issue which Lisney placed about the year 1814.

It was not unusual practice for natural history works to be published at that time in French as well as English, a procedure undoubtedly calculated to attract continental buyers: for example *Gleanings of Natural History* by George Edwards, which was published in

this copy of the second issue, but does not include the three additional coloured plates and the anatomical diagram, present in the copy before us and in all subsequent issues and editions. Anxious to extend the first issue to portray some additional species of lepidoptera, Harris decided to produce a Supplement, including these further plates and adding also the Table of Terms and Index. There appears to be no record of this Supplement as a separate publication in existence today, although Lisney records a version with four plates, which seems to be unique; they do not appear in any edition of *The Aurelian*.

In due course these supplementary plates and accompanying text were incorporated into the second issue which was published about 1773, although if the title page is referred to, it will be seen that it is the original one, dated 1766, of the first issue; the last plate in this second issue is dated 1 March 1773.

This approximate publication date can be arrived at by reference to an advertisement in an excellent little handbook which Harris published in 1775, entitled *The English Lepidoptera: or, The Aurelian's Pocket Companion*. This little book of forty-two leaves, with a coloured anatomical plate of a butterfly, was the first book on English lepidoptera to give a great deal of information, such as times of appearance, food plants and guidance on collecting and breeding — all of which was very valuable advice to the practical collector: upwards of four hundred species were catalogued, giving the English and Linnean names where known. From the Preface to the work, we learn that contained in it are the results of many years careful observations, the commencement reading as follows:
'The usefulness of the following little work will be obvious at first view, to anyone who has the slightest acquaintance with this part of Natural History: The author found it so necessary, that he always carried a copy of it in his pocket ever since he began to collect the different species of Lepidoptera.

What is now offered to the Publick has been revised and properly arranged, so that it is truly a compendium and repository of every new discovery of the author's researches for almost these thirty years diligent application.'

three parts (1758-1764), and James Barbut's *Genera Insectorum of Linnaeus* (1781) adopted this style. Moses Harris's earlier promise in the *Proposals* that a French translation would be given with each part was not fulfilled: he quickly realized the impossibility of this, for on the cover of the third fascicule (1759) in Lisney's possession appeared the following note, probably inserted by the publishers:

'N.B. to Prevent any future Delay in the Publication the french Translation will Be Delivered Occasionally.'

Six years after Harris's death a third edition was published 'imprimé pour J. Edwards 1794', with a French title page and the English and French texts printed, not in double columns, but on opposite sides of the leaves. In the same year or shortly afterwards another issue appeared, similar to the first, but with an English title in lieu of the French.

Finally, a *New (4th) Edition* edited by J. O. Westwood was published by Henry G. Bohn in 1840, for which proof copies were issued in the previous year. All these later editions had the same number of plates, and in some cases the difference in the contents was slight.

So much for *The Aurelian*, which from an artistic viewpoint is without doubt unsurpassed in the literature of British entomology, and anyone fortunate enough to possess a copy must give it pride of place in the library. Moses Harris rarely copied other illustrators, but his insects were drawn and coloured *ad vivum*, usually from specimens collected on his frequent field excursions, which appeared to be confined to a large extent to the counties bordering the Thames estuary.

Works of this kind with coloured illustrations were frequently published in a cheaper edition with the plates uncoloured, but no such copy has yet been recorded, although a copy with plain plates was acquired, probably from Harris, by a contemporary entomologist, Henry Seymer. This he coloured himself, for his own satisfaction.

Many painters in the exercise of their art became attracted and intrigued by the sometimes arresting and often delicate and subtle colouring of insects. One such was the well-known teacher of painting and drawing, Eleazer Albin, who disclosed in the preface to his *Natural History of English Insects* (1720) that, 'as the various forms and colours of flowers and insects gave him great pleasure a deep interest developed which led him to study them more closely'.
This was the first volume to be published in England devoted entirely to insects, illustrated with hand-painted plates, mostly of butterflies and moths.

Others included Benjamin Wilkes, mentioned earlier, a talented painter of miniatures, also a member of the first Aurelian Society; and Joseph Dandridge, an outstanding entomological artist (although not so well known, as nothing was published under his own name) who was also an Aurelian.

It was Moses Harris's close association with entomologists from an early age, particularly with his aurelian uncle of the same name, which must have left him especially aware of the tremendous variety of form and colour in the insect world — an appreciation which later produced such splendid results: a little of that appreciation can be experienced today through the medium of natural-history films on colour television.

The strange new world must have been enchanting to Moses Harris, whose particular attention to colours made him pre-eminent as an illustrator. The famous nineteenth-century naturalist, William Swainson, pronounced Harris to be 'the best painter and engraver of insects of his day, besides being a most accurate describer'.

A unique addition to the illustrations in Harris's *An Exposition of English Insects* (to be mentioned later) was a plate of 'explanation of colours' in the form of a circle with different colours radiating from triangles of colours in the centre, with explanatory text. His *Natural System of Colours*, which was edited by Thomas Martyn and dedicated to Sir Joshua Reynolds, was published in London in 1811, long after Harris's death. An adaption of this by Turner, viz. *Colour Design No.2*, hangs in the Turner Gallery of the Tate Gallery in London. In the list of major artists who exhibited at the Royal Academy during his lifetime, the name of Moses Harris appears but once, in 1785, when a frame of insects was hung there.

The finest period in English history for the flowering of the arts was the Georgian era when, despite the Seven Years War and the War of American Independence, the social conditions were favourable for the production of fine works of art, 'Life and art were still human, not mechanical, and quality still counted far more than quantity' wrote Trevelyan of this era in *English Social History*. Eighteenth-century England was flourishing under the leadership of the 'social aristocracy of that day (which) included not only the great nobles but the squires, the wealthier clergy and the cultivated middle class who consorted with them on familiar terms'.

And so it was to this class that Moses Harris looked for patronage. The reader will see at the foot of each plate (except for numbers five and thirty-four, the three which were added to the first issue and the anatomical plate) a Dedication which perpetuates the name of the patron who helped, by a substantial subscription, to defray the expense of the plate. This practice was frequently adopted by authors of illustrated works — it pleased the patron who was anxious to be known as a generous supporter of the arts and learning, and it enabled and encouraged the author to produce a work worthy of his art. Many volumes thus found their way into the private libraries being formed by an increasing number of wealthy people at that time.

Five guineas was the cost of *The Aurelian* — a reasonable price for a book of this sort in the middle of the eighteenth century, but still enough to put it beyond the reach of all except the well-to-do. Probably because of its entomological significance and importance for reference, several interested people clubbed together to purchase a copy, and the fact that it was issued in parts, at intervals, helped to increase the sale. It must be admitted, however, that if the bibliographers are right, only a portion of the book was published in parts, the balance being issued and paid for together, thus necessitating a considerable outlay. Disagreement and delays were, it seems responsible for this unpropitious birth.

Harris's next publication *An Exposition of English Insects* was of great scientific importance; in this he showed himself as a serious taxonomist, being responsible for the naming of many species of diptera and hymenoptera and delineating and colouring several orders of insects besides lepidoptera. The book was produced in quarto and so had smaller sized leaves than *The Aurelian*, and it lacks the impressive aesthetic appeal possessed by the composition of the plates in *The Aurelian* — which portray also flowers and plants, as well as the paraphenalia of collecting that are found so particularly fascinating by the modern entomologist.

It was published in five decads, the first in 1776 and, according to Lisney who had carefully explored the bibliography of the book, the second decad may also have been published separately. But the last three he is sure were all published together, probably in 1780. Over a period of ten years there were published two editions, with two issues to each; the first edition having the text in English and French in double columns. The first issue is particularly celebrated as this contains the engraved self-portrait of Moses Harris which he signed 'Moss Harris delt et sculpt 1780. Done at the desire and expense of Mr. I. Millan'. The portrait is the only one recorded of Harris, and is surrounded by attractive engravings of his 'tools of trade', butterfly nets, a palette, some books, scrolls, a pipe and so on, as well as the inevitable lepidoptera and flowers. It is inscribed 'Moses Harris Aeta 49', from which the inference was drawn by biographers that he was born in 1731, now known to be 1730, as mentioned earlier.

The attractive frontispiece to *The Aurelian* is of a silvan scene with a seated figure of a fashionably dressed young man in the foreground; he holds his 'bat-folder' net and trays of insects, with his right arm flung out as if declaiming the text at the foot of the engraving: 'The Works of the Lord are Great, Sought out by all them that have Pleasure therein. Ps CXl.v.2'. It is not certain, but this figure in an arcadian setting is generally thought to be also a self-portrait of Moses Harris. Certainly there does appear to be some similarity with the signed portrait which is published in *English Insects*. A little further in the background is another figure, similarly dressed, engaged in collecting insects with a net in a woodland glade.

That his services as an artist were much in demand is evidenced by the commission of fellow authors to illustrate their works. The most important of these was the three-volume *Illustrations of Natural History* (1770), written by Dru Drury, a Strand silver and goldsmith, and a patron and friend of Harris, with whom he frequently went on collecting expeditions. Harris dedicated plate thirty-seven in *The Aurelian* to him. Each volume of Drury's magnificent work contains fifty coloured engravings, mostly the work of Moses Harris, with the accompanying text in English and French.

A huge collection of insects was built up in due course by Drury, contributed to chiefly by his many correspondents who lived or travelled abroad. Thus many strange and wonderful species rarely, if ever, seen before in Britain were brought or sent home at his request. He was a man of great integrity who paid fair prices for them; but that price, as may be imagined, was sometimes difficult to decide, particularly if the sender wished to leave it to Drury's generosity. On one such occasion, when writing in 1764 to a business man in Bombay and faced with this dilemma, he wrote, in an attempt to be quite just, 'If you approve of it we will leave the price to be settled by ye Aurelian Society'.

Very careful printed instructions for the use of foreign collectors were given or sent to them offering advice, such as what to look for, how to pack the specimens, and the like. Nothing could be more exasperating to Drury than to receive broken or damaged insects, particularly after a long wait for a reply, and the lengthy voyages in the slow and uncertain days of sail presented further hazards. In a letter to a correspondent in India (written in 1762) he showed justifiable annoyance when he wrote, 'Your letter and ye box of Flys came to hand but the Flys were all of them spoilt, not one of them being left whole as you may see by the remains wch I have sent back to you, it was a great pity, you cannot think how we were vexd. about it'.

It was an age when the tempo of life was so much slower, yet Drury's patience must often have been sorely tried, particularly on one occasion, when he was anxious that publication of his *Illustrations* should not be held up, and so took Moses Harris severely to task for being too leisurely in completing certain plates. The drawings, engravings and best copies of the plates were all executed by Harris, 'who', said the author of the sympathetic *Memoir* of Drury, in Jardine's *Naturalist's Library*, 'is admitted by all to have been one of the best, if not the very best, entomological artists of his day. His employer also exercised a vigilant superintendence over his proceedings, so that the utmost confidence may be placed in the general accuracy of the delineations: both in regard to engraving and colouring. Indeed, the care with which they were got up, seems occasionally to have led to such delay as nearly to exhaust the patience of the placable author and disturb his habitual equanimity'.

The reason for Drury's anxiety was that some sets of prints were collected from him by Moses Harris's son on February 8 1770, and had not been returned duly coloured some two months later, apparently long after the expected date. Drury was very worried, not least because of the adverse effect on the sales of his book then in the process of being issued in part numbers. It is not known where Harris was living at the time, but in Drury's letter to him of April 5 of that year, he wrote, 'I wish

to Heaven you was removed from the place where you are now buried, and come to London, for then I could scold you by word of mouth, and am now forced to employ a great deal of time in doing it by letter, which I can but ill spare. I beseech you, if you have not heard of a house in London, put an advertisement in the newspapers for one, that I may not be compelled to put you and myself to the trouble of reading and writing these letters'. Despite his irritation which on this occasion bubbled to the surface, Drury knew he couldn't do without Harris, such was his reputation.

It is interesting to note that of his *Illustrations* it is recorded that 'two series were issued at different prices, one of them wholly coloured by Harris's own hand, and the other by ordinary colourists under that artist's superintendence'.

Drury felt obliged after a severe financial reverse (of which, alas, he had several in his lifetime) to take steps to sell all his collections and drawings. To this end he wrote to the influential Dr Pallas, the German naturalist and traveller who spent many years in Russia, with the suggestion that his friend the Empress of Russia might prove a purchaser. In his letter he wrote, 'I also intend to dispose of all the original drawings from which the engravings have been taken. These, I must inform you, are wonderfully fine — Mr. Harris, the painter, who executed them, and also engraved and coloured my work, having exerted his utmost skill in making these drawings. The latter I value at £325'. Harris must surely have been quickly forgiven for his tardiness! In fact, Drury did not proceed with the sale, which took place in 1805, about a year after his death.

The spirit of genuine scientific enquiry was engendered by the enthusiasm of such indefatigable collectors as Drury and Wilkes, as well as Petiver (1663-1718) and others of an earlier generation, and so the need arose for a rather more comprehensive guide for the use of travellers than the printed broadsheets distributed hitherto. The genial William Curtis (1746-99) brought out a little work, of great value to entomologists, entitled *Instructions for Collecting and Preserving Insects* (1771), which included an engraved plate showing the way insects were to be pinned and set, and collecting apparatus.

These entomological methods, described in such an interesting way by Harris in the Preface and elsewhere in *The Aurelian*, present fascinating reading for the modern collector, especially when compared with those that are used today.

One of the best-known English naturalists, Curtis is perhaps more appreciated for *Flora Londinensis* (1775-98) and for the foundation in 1787 of the *Botanical Magazine*, the unrivalled serial publication still continuing, but he was also a fine entomologist.

After Curtis's death, a number of drawings of insects by Moses Harris were found in his possession, which were among the many treasures housed in the Curtis Museum at Alton in Hampshire. This is a museum

founded in 1855 by a kinsman, William Curtis, in the town where his famous forebear was born. In a biographical account of the great botanist, published in 1941, W. Hugh Curtis concluded that these drawings were intended to provide illustrations for a natural history of the British Isles: this would have been a publication of considerable importance, but for some reason it was not proceeded with. We must remember that Curtis was only fifty-three when he died in 1799. Confirmation of his intention came a few years after his death when A. H. Haworth, in the preface to his now rare *Lepidoptera Britannica* (1803), referred to his 'late ingenious friend Mr. William Curtis' who had the idea of forming a society 'to collect and explain all the natural products of Great Britain, or at least, the entomological ones' but, said Haworth, Curtis abandoned the idea and passed his collection of British insects to him with encouragement to undertake this great work.

Haworth therefore had a good start when in 1801 he formed a new Aurelian Society and was appointed the first Curator of its cabinet of insects. Curtis suffered from heart trouble for several years, so he may well have realized his ambitions would never be fulfilled, although it must be said that, in spite of his outstanding successes, he earned the following comment from his friend Samuel Goodenough, who in penning his obituary notice, wrote; 'Mr. Curtis was perpetually forming some new design without completing any one'. A forgivable trait in a man of such generous disposition and enquiring mind, living in an age of so many new discoveries. We are fortunate that these original drawings have survived. They are now deposited with the Hampshire County Museum Service in Winchester.

The year following Curtis's publication of the *Instructions* saw another work, rather more extensive, written by a fellow Quaker, a doctor named John Lettsom, who was a lifelong friend. Indeed, it was the philanthropic Lettsom (1744-1815) who gave substantial financial support to Curtis when the poor sale of his superb *Flora Londinensis* was causing him much worry. The title of this book was *The Naturalist's and Traveller's Companion*, much broader in scope than Curtis's, for not only were instructions given for discovering and preserving insects, but it contained information on many other natural-history objects such as seeds, fossils and so on. Also included were 'Directions for taking off Impressions or Casts from Medals or Coins'! As may be imagined, Lettsom was himself a keen collector with wide interests. The frontispiece of this guide, (which clearly fulfilled a need, for it passed through three editions, with preparations for a fourth) was a coloured plate drawn and engraved by Moses Harris, illustrating a way to pin insects.

Towards the end of his life, Harris also undertook some drawings of plates for a fine two-volume work by William Martyn, entitled *New Dictionary of Natural History*, published, according to the title-page, in 1785; of the total of one hundred engraved and coloured plates, however, only nine are actually signed 'Moses Harris del et sculp'.

These are mainly illustrations of butterflies and moths but it is likely, although there is no proof, that he was responsible for the remainder. Of the author himself, nothing, besides this work, is known. Lisney draws attention to the fact that the book may have been issued in fascicules, as the plates are dated at weekly intervals, the last one being 19 May 1787. Of the hundred plates, the latest signed by Harris (who died about the year 1788) was numbered ninety-seven.

A few original paintings by Harris have found their way into important institutions, although probably more have survived unrecorded, finding a place in private ownership. Deposited in the British Museum (Natural History) are some original watercolour paintings for *Historia Naturalis Testaceorum Britanniae* — Emanuel Mendes da Costa's well-known conchological work, which was published in London in 1778. Included in the collection is one by Moses Harris, for whom da Costa had a high regard, describing him as a 'famous entomologist and miniature painter'. There is also a watercolour on vellum of a tarantula, to illustrate Henry Smeathman's *Account of the Tarantula of Sierra Leone* and signed 'Moses Harris pinx 1767', which is deposited in the Library of the Linnean Society in London. Smeathman, who was a professional collector operating in West Africa, particularly Sierra Leone, was a member of the Aurelian Society of which Harris was Secretary, and a close friend of Drury to whom he sent numerous specimens. In fact, Drury wanted Smeathman to write the last volume of his *Illustrations*: he did not do so, but his extensive notes on the specimens he sent from Africa enabled Drury to complete it.

It will be noticed that in the brief commentaries opposite *The Aurelian* text which accompanies each plate, the modern scientific name has been indicated, according to that published in *A Recorder's Log Book of British Butterflies and Moths* (1979), by J. D. Bradley and D. S. Fletcher. Scientific binomial nomenclature was not in general use when *The Aurelian* was first published, although used (probably for the first time by an English author) in John Berkenhout's *Outlines of Natural History of Great Britain and Ireland* published in 1769.

This followed the publication in 1758 by the great Swedish naturalist, Linnaeus, of the tenth edition of his *Systema Naturae*, first published just over twenty years earlier and now adopted as the starting point for zoological nomenclature. Latin, as well as English names, were used by Harris in the *English Lepidoptera* in 1775, and many Linnean specific (trivial) names are given in the supplementary additional leaves in the copy of *The Aurelian* before us. They are, in fact, taken from the later twelfth edition of *Systema Naturae*.

Most of the butterflies and quite a few of the larger moths illustrated by Harris were first described with accompanying crude woodcuts in Thomas Moffet's *Insectorum sive Minimorum Animalium Theatrum*, printed in 1634. Although some species in it are difficult to recognize, much of this uncertainty has been dispersed by Canon Charles Raven in his masterly study of Moffet's book in *English Naturalists from Neckham to Ray* published in 1947 by the Cambridge University Press.

We owe a great debt to Harris and his circle for the legacy they left us and the groundwork of knowledge upon which we, of a much later generation, are still building. But Harris did more than that; he left us a book the like of which we shall not see again in this age of highly specialized entomological professionalism. Should interest in our native butterflies and moths be stimulated by the beautiful paintings in *The Aurelian*, there are many excellent books on the subject published today. Moreover, there are few areas of the country in which natural-history or conservation societies have not been formed, and from which enquirers may obtain advice and a welcome to join in field-work. So much easier today to join the 'brotherhood of the net' than in the days of the earlier struggling Aurelian Societies, confined just to the metropolis and those of sufficient sagacity!

However, amassing a collection will not make one a naturalist. Indeed, unless for a very good scientific reason, it is not a practice to be encouraged. Rather should Harris be emulated in such things as breeding experiments, in which his keen observations were of great value; even today, little is known about the life history of some of our British species, so there is still much scope for the amateur entomologist to achieve some worthwhile discovery.

The lacunae in our knowledge of the lives and accomplishments of many of those eighteenth-century and the earlier enthusiasts provide many opportunities for the natural historian to put the breath of life and colour into an image which is so often little more than a name with a sketchy profile.

The entire quaint text of *The Aurelian* has been reproduced with the plates but, for simplicity, Harris's long 'f' has been replaced by the modern typographic 's'.
Where the first reference to an early author is made, the date and title are also given but this has not subsequently been repeated in full.

Acknowledgements

Since the far-off days of the publication of this historic pioneering work, entomological knowledge has progressed apace; it must be borne in mind that most of Harris's comments were the result of firsthand observation, in which he displayed a patience rivalling that of the great French naturalist, Jean Henri Fabre. Among the most useful of the reference books now available to the enthusiast are the *Colour Identification Guide to Moths of the British Isles* by Bernard Skinner (1984), which includes all the larger British moths, and the *Colour Identification Guide to the Butterflies of the British Isles* by T. G. Howarth (1984): both give comprehensive information. They have, *inter alia*, been consulted, but the abbreviated notes opposite Harris's text are for general interest and guidance only; they are not intended to replace the detailed information which these books provide. Throughout this account of *The Aurelian* a number of authors have been mentioned, and in addition to the works cited, the historical and bibliographical essays which have appeared in the *Entomologist's Record* and the *Entomologist's Gazette* over the last decade or so have been a source of great interest and illumination.

My warm thanks are due to many people for their help, in particular to David Wilson for his interest and encouragement and to Maitland Emmet MBE, whose knowledge of the 'smaller lepidopterous tribes' is unrivalled; I am greatly indebted to Michael Chalmers-Hunt who undertook the valuable but unrewarding task of reading and checking the typescript.

The responsibility for any errors is mine entirely; I have no wish, in the words of the illustrious Harris 'to propose to myself . . . to extenuate any Neglect of my own'.

DEDICATION. TO THE PRESIDENT,
And the Rest of the GENTLEMEN,
THE WORTHY MEMBERS of the AURELIAN SOCIETY

GENTLEMEN,

IT is in Gratitude to the great Friendship and Encouragement this Virgin Volume of Mine has met with at your Hands, which first prompted me with a Desire to lay it before you; not intending it as a Compliment to recompense Favour, but to shew my Respect and Esteem to our Worthy President, and my Brother *Aurelians.*

It is to your Protection then, Gentlemen, that I submit this Volume, poor in Worth, hoping for your favourable Construction on the many Disappointments You and the World have met with in the Course of this Work, and the tedious Length of Time it has been compleating; neither can I in Justice to myself, say the Fault has been wholly my own, whose earnest Desires and Endeavours, it has been continually to forward and compleat it; but owing to the unsteady and fallacious Behaviour of a Person, too nearly connected in my Concerns: Neither do I propose to myself, by complaining of his ill Treatment, to extenuate any Neglect of my own.

I shall, Gentlemen, in the next Place, take hold on this favourable Opportunity, to assure you of my constant Adherence in endeavouring to promote the Interest of the Society; and to assure you of my Love and Friendship for every of you, whose Regard I hold much in Esteem, wishing you Success in your praise-worthy Pursuit, not only in Collecting, but carefully considering the excellent Works of the Almighty and Supream Creator, which he has pronounced to be good: For my own Part, I shall still persevere in this my beloved Employment, and hope, if God permit, in a short Time to produce to you and the Publick, a farther Account of our *English* Insects, in which I hope for your farther Assistance and Encouragement.
I remain,

Gentlemen, In all Sincerity, Your Most Humble, and Obliged Servant,

MOSES HARRIS.

PREFACE

IT is now above twenty Years since I first began to collect and pursue the Study of Insects; the first Hint I received was from Mr. *Moses Harris,* an Uncle of mine, who was then a Member of the old Society of Aurelians, which was held at the *Swan Tavern,* in *Change-Alley: I was then too young to be admitted a Member tho' the strong Inclination I had to be searching into this Part of Natural History, made me very desirous: I was then but twelve Years old, so obliged to defer it, till Age should ripen and furnish me with sufficient Sagacity, whereby I might become fitting for the Company of that ingenious and curious Body of People. I was, however, deprived of that Pleasure, for not long after the great Fire happened in Cornhill,* in which the *Swan Tavern* was burnt down, together with the Society's valuable Collection of Insects, Books, *&c.* and all their Regalia: The Society was then sitting, yet so sudden and rapid was the impetuous Course of the Fire, that the Flames beat against the Windows, before they could well get out of the Room, many of them leaving their Hats and Canes; their Loss so much disheartened them, that altho' they several Times met for that Purpose, they never could collect so many together, as would be sufficient to form a Society, so that for fourteen Years, and upward, there was no Meeting of that Sort, till Phoenix-like our present Society arose out of Ashes of the Old. However, my Fondness for Insects made me very industrious, and I have continually endeavour'd to take all Opportunities, to get Knowledge in the Times, Seasons, and Manner of breeding them; Part of which Experience, join'd with the Assistance of my Friends, I do now, with all Submission, lay before the Publick.

It is likely the Reader will find many Things in this Work, that may not meet with his Approbation; but as I never before attempted to engrave, I hope the Faults which may be found in the Plates will be excused on that Account; my Manuscripts was at first setting out, for the first three Pages, corrected and put into Form by a worthy Gentleman of my Acquaintance, Mr. *Walter Wall,* by whose Intreaty I published this Book of Insects. But his Affairs calling him to the *East-Indies,* I lost at once an ingenious Friend, and a kind Assistant to this Work: Thus being left to myself, my Manuscripts went rough from my Hand to the Press, bare of those pretty Embellishments with which Men of a more liberal Education sprinkle and adorn their Writings, making them pleasing as well as edifying; but if mine may

want what a more learned Hand could bestow, yet what I have wrote is Truth, which is ever most Beautiful when most exposed, so if I can but be understood, it will, however, have two of the most principal Embellishments intended to adorn it, viz. Truth, with Instruction. But enough of this; for as I have much to say, I shall desire to be short with that Part I count the most immaterial, and proceed to the Marrow of the Matter in Hand, I mean that of the Introduction: And First.

That every Caterpillar produces either a Moth or a Butterfly, Nobody that has had ever so little Insight into this Part of the Creation, will venture to dispute; I mean such Caterpillars as have not less than ten or more than sixteen Legs; for those which have more, produce what we have hitherto distinguished by the Name of *Ichneumons,* and such as have less commonly produce the Beetle or Chaffer Kind. But as I am here about to display their Proceedings, under each of their several Changes or Appearances, wherein they are considered in a General Light. I shall first begin with the Egg, taking their other Appearances in their natural Order as they follow; thinking by thus explicating on the Subject in a general Manner, the better to prepare the young AURELIAN for the Perusal of the following Work, where each Insect, whose History is known, is particularly considered and represented in its different States.

The Females, both of Moths and Butterflies, lay their Eggs in a few Hours after Copulation, upon or contiguous to what is design'd to be the Food for the young Caterpillars; when they appear from the Shells, some produce the Caterpillars in fourteen Days, others in four or five Weeks, some again do not appear till the Expiration of four or five Months. When the young Caterpillars are perfect within the Shells, they eat, or rather crumble the Shell away with their Chaps, and feed on what was provided by the Parent for them; and each of these Caterpillars having grown to their full Size, and purged themselves from their Dung and Filth, cast off their last Skin and becomes a Chrysalis, or an Aurelia; from which a Fly or Moth is produced in the Likeness of its Parent, after lying in that State a certain Number of Days, Weeks or Months, according to the respective Species or Class to which they belong, and this is their general Progression in which Particularities they do both agree.

Of the Eggs, and the different Manner of Laying them.

THE different Classes of both Moths and Butterflies, deposite their Eggs remarkably different, some fasten them to the Food by a viscous Moisture, detached from each other, a small but irregular Distance; others lay them in a confused Heap, fastened together like a Lump of Sand; some range them in regular Order like a curious Pavement of round Pebbles, these are on a plain or flat Surface:

But some others lay them round a Stalk or Blade of Grass; some cover them with a woolly or downy Substance, which keeps them from the Cold, and hides them from the Sight of Birds; others drop them, as they fly, loose on the Ground in a promiscuous Manner, of these most of their Caterpillars feed on the Grass.

On the Caterpillar, and its Change to the Chrysalis

In their Progress from the Egg to the Chrysalis, they shift or throw off several Skins generally one every seven Days; but I am not certain whether they all shift the same Number of Skins alike; Those which I have noticed have cast their Skins five Times, and 'tis supposed they all do the same; however, they are about seven Weeks in their Caterpillar State; at the Expiration of which Time they are full fed, and prepare for a future State, by making themselves a secure Retreat, wherein they lie two or three Days, during which Time they shrink and grow shorter, losing the Use of their Feet entirely, and appear as if in great Agony; at Length the Skin on the two first Joints behind the Head, which at this Time appear very much swell'd, bursts, or rather splits, and opens some Way down the Back, and cross the Head, that in some you would at first Sight suppose the Head of itself was divided. During this Time the Caterpillar strives to throw off its Skin, which however he facilitates by a Motion very peculiar, working off the Skin Joint by Joint, till it arrives quite to the Tail; nor does it cease twisting and turning itself till quite disengaged from it: It is now very soft and tender, and it is generally a Day before the Shell of the Chrysalis becomes hard, during which Time of hardening it very frequently turns itself, that the Side on which it lies may not be flatulated or deformed, yet when the Shell becomes hard, it lies Motionless, unless disturbed by some Accident, till the Expiration of a certain Time, and at length breaks forth in the winged State; but during the Time of its being in the chrysalis State, it receiveth no Nourishment of any Kind, altho' some remain in that State near two Years.

From the Chrysalis to the Fly.

When the Fly is perfectly form'd within the Chrysalis, or more properly, when each Part has arrived to its proper Shape, Strength, and Texture, the Chrysalis then appears much darker; and if a Butterfly, the Markings of the Wings plainly are seen through the transparent Chrysalis; at this Time the Hull, or Shell of the Chrysalis, is separated from the Fly, whose every Part begins to grow dryer, whereby it is the better enabled to separate them; thus being as it were unbound, and capable of moving, it makes a strong Effort at once with that Part which I shall call its Shoulders, and pushing at the same Time with its Legs forward, it splits the Head-part of the Shell in three Divisions one in the Front, which covers its Legs and Face, the other two one on each Side, covering the Wings; it then bends itself forward, and the front Division or Mask yielding, it lays hold on the lower or jointed Part of the Shell, and draws itself intirely out; being disengaged from the Chrysalis, its next Business is where it can hang by its Legs, with its Wings downward, and where they may stretch and grow without Obstruction. For the Wings of the largest Flies, when they first come out of the Chrysalis, are not much bigger than a silver Penny.

It seems very careful in the Management of its Wings while they are growing, often shaking its Body, least by their Dampness they should stick together; then by a gentle rocking of its Body, to try if it can feel with its Wings any thing that may obstruct their growing; and if so, will directly creep higher or move at a greater Distance. As the Wings grow they rumple and pucker in rude Fashion, and after a short Time, they nicely expand themselves, hanging very flat, and exactly even with each other: They now appear of the Consistence of Paper which is rotten damp; but they are much less Time in drying than in growing to the full Size. When he thinks his Wings are ready, he suddenly opens them a little Way, and if he finds them too heavy, or that they yield, bend, or give Way in striking the Air, he very cautiously closes them again; at length he begins to open and shut them softly, as it were fanning them lightly, till they be quite dry, then suddenly starts into the Air, and flies away.

A Butterfly is fit for Flight in less than half an Hour after it comes from the Chrysalis, and those of the greatest Moths hardly exceed an Hour.

The Food of both Moths and Butterflies are much alike, being chiefly the Honey which they extract with their long Proboses, out of Flowers, or the Honey-dew which is found on the Leaves of Trees, Plants, &c. Indeed it may be excepted, that some Moths do not feed, nor take any Sustenance whatever, the Hens in particular, nor have some any visible Organs for such Purpose.

How many different Species of Moths and Butterflies we have now in England, is certainly impossible, even to form a Judgment of: However, of those already known, our Catalogue amounts to between four and five Hundred, of Flies we have about Fifty; neither is there much Expectation of there ever being any more discover'd: But of Moths something new may be found almost every Day, if sought after with Diligence.

The Distinguishing Difference, between a Moth and a Butterfly, considered in their Caterpillar Chrysalis, and Winged State.

First then, to begin with the winged State. The Horns or Antenna of a Butterfly, hath a Knob, or Ball at the Extremity or End of each, and are for the most Part pretty straight. The Antenna of Moths, chiefly diminish gradually, and end in a sharp Point; tho' indeed some do swell about the Middle, or towards the Point, or End, some are Comb-like and very broad, others appear like fine Thread; and most of them have a winding Motion from the Root to the Extremity; others are notched on the under or inner Side, like the Teeth of a Saw.

The Tails or Abdomen of Butterflies, commonly lays in a Kind of Grove or Bed, which is form'd by the Underwings; neither do the Tails ever reach below the Edge of the Underwing. The Tails of Moths in general, lie beneath the Underwings, and reach to the Extremity of the Underwings, and many of them a great way beyond.

A Butterfly always sleeps or rests with its Wings erect over its back; the Underwings being Broad, and without Folds. The Moth commonly rests with its Wings covering its Tail and Underwings; on which Account, Providence has ordered it so, that the Underwings of all Moths, fold themselves up half Way in the Manner of a Fan. A Butterfly always flies in the Day, and mostly in the Morning, but never in the Night. Moths, some flie in the Day-time, some after Sun sett in the Evening, and others in the Dead of the Night.

Tho' I must here observe, that tho' these are Rules very sufficient whereby you may know a Moth from a Butterfly, yet I well know there is not one of the above Rules, but has an Objection; and altho' Moths and Butterflies are two quite different Species of Insects, and at the very first Sight each of them confess themselves to a good AURELIAN; Yet I wholly believe 'tis impossible, by any proposed, particularily, to make a Rule to know one by, which some one of the other will not be an Objection to. The same Difficulties arise in attempting to class either the Moths, or the Butterflies; and altho' several has attempted to do it, yet the many Obstacles they meet with in their Way, especially among the Moths, many of which will jar, with the Order and Regularity of their Work, that the Matter is rendered extreamly Difficult; these however, they squeeze into some Place or other; for you know they must be put some-where; so at last the Work is rendered Imperfect; but to return to the Matter in Hand.

I shall further observe, that when a Moth in any Part of its Investure seem to close with a Butterfly, viz either in the Horns, Shape of the Wings, &c. it shall differ in every other Particular, and that so notoriously, that you may always be able to know which of the two it is. Thus the great Magpie in the Form of its Wings much resembles a Butterfly; but its Horns or Antenna, are like crooked Threads; the Head remarkably small, its Abdomen reaches to the bottom Edge of the Underwings, which have several folds in them; and as all this Class of Moths do most resemble a Butterfly at first Sight, so the Caterpillar from whence they proceed, has the least Resemblance to those of the Butterfly Kind; for as I have often observ'd, those broad-wing'd Moths, mostly proceed from Loopers, which have the smallest Number of Legs, while those of the Butterfly Kind have the most; and indeed, not to be too prolix, I know not any other Matter, either Texture of Wings, Manner of Flight, Food, or any Thing else in which it any Way resembles the Butterfly Kind. This brings to my Mind, Mr. *Wilks's* new Class of Moths, *Book* II. *Chap.* 1. where I think he says, Flies resembling partly Moths, and partly Butterflies; here he produces but one single Instance, which to me is no Instance at all of any such strange Being; nor would I have taken the Liberty to have mention'd it in this Work, had I not been fearful young AURELIANS might have been led astray, by the Novelty of the Thought, and perhaps in Process of Time, have added a great Number of other Moths and Butterflies into a Species of Insects, never before heard of, at least in this Part of the World. The Fly he speaks off, is the Burnett, the Horns or Antenna of which swell, and that pretty much towards the Point or Extremity: But this is not at all like the Ball or Knob at the End of the Butterflies Antenna; in its setting Position, its Wings are remarkably closed about its Body, which appears much below the Underwings; but indeed this Insect in every Circumstance least resembles a Butterfly, than any Genus of the Moth Kind that I know, either in the Form or in the Colour, which in this Moth is so very remarkable, that none of our Butterflies has the like; I mean that fine shining Metallick Green, which covers the Body and upper Wings of this Moth; and where Greens are introduced in the Butterfly Kind, I mean only these of this Island, you will always find it on the under Side, and not a Green of this shining Quality; but to re-assume my former Discourse.

Their Caterpillars likewise differ in the following Particulars.

Those of the Butterfly Kind have all in general, that are yet known, sixteen Legs, and those placed in the Manner following: Supposing that Caterpillars of all Kinds have twelve Joints, Rings, or Divisions, where the Caterpillar's Legs are placed is marked by Dots, counting from the Head to the Tail. The First toward the Head are in the Form of Hooks, or Claws; the Eight which are in the Middle, may with more Propriety be called Legs.

1	2	3	4	5	6	7	8	9	10	11	12
0	0	0			0	0	0	0			0
0	0	0			0	0	0	0			0

Being form'd with the same fleshy Substance with the Body, and is most employed when the Animal is in Motion; the two last behind may be call'd Holders, because with them they do hold or adhere very strongly to the Leaves and Branches, nor do they ever loose them when creeping, till all the rest of the Feet are fixed.

The Caterpillars of the Moth Kind differ greatly with regard to the Number of Legs, which I shall divide into five different Classes, *viz.* Loopers, Half Loopers, Quarter Loopers, those having no Holders behind; and the common Sort, which have the same Number with those of the Butterflies; the Legs of the Loopers are placed thus:

1	2	3	4	5	6	7	8	9	10	11	12
0	0	0						0			0
0	0	0						0			0

The above move forward by stretching or extending themselves at full Length, holding fast at the same Time with their hind Holders, and the next you see placed at (9); they then fix fast their six Claws at the Head, and loosing their Holders, draw their Tail-part quite close to their Fore-part, so that at every Step or Stride they form a Loop, and for that Reason are called Loopers. The Legs of the Second are placed thus, and are called Half Loopers; because in walking or creeping, they bend their Bodies in the Form of a half Loop; or to speak more intelligible, the Legs of the Hinder never are drawn up close to the Legs of the Head-part, which the Loopers always do.

1	2	3	4	5	6	7	8	9	10	11	12
0	0	0					0	0			0
0	0	0					0	0			0

The Third hath fourteen Legs, which are placed as beneath, and are for Distinction Sake called Quarter Loopers, and with the same Propriety as the Last are called Half Loopers; for they bend their Bodies in the Form of Part of a Loop, tho' not so much as the Half Loopers do. The Legs of this Sort are thus placed:

1	2	3	4	5	6	7	8	9	10	11	12
0	0	0				0	0	0			0
0	0	0				0	0	0			0

The Fourth hath likewise fourteen Legs, tho' placed in a different Manner, they having no hind Holders.

1	2	3	4	5	6	7	8	9	10	11	12
0	0	0			0	0	0	0			
0	0	0			0	0	0	0			

The Legs of the Fifth are the same with the Butterflies, both in Situation and in Number, which is Sixteen, nor has any Caterpillar more.

The Caterpillars producing Butterflies do not seem very inclinable to be hairy, some being quite naked, others cover'd with a short woolly Down, something like that upon the Peach, the rest are beset with branched Spikes, which those of the Moth Kind never are; those Caterpillars of the Fly Kind, which may be said to be hairy, the Hair is very fine and tender; of those of the Moths, some are thickly cover'd with long Hairs, others have very few but those long, some have Tusks or Tussocks of Hair on their Back, the rest are quite naked.

The Caterpillars of the Butterfly Kind, when ripe for their Transformation, some hang up by the Tail, with their Head perpendicularly downward, which are those of the thorny or spiked Kind; and one other Class of which we have but one Specimen, *viz.* the Purple Emperor, with this Difference between the two, that the one always choosing the under Part of an horizontal Plane, such as the Ceiling of a Room to hang from, the other from a Perpendicular, such as a Wall, &c. the others fix themselves by the Tail with their Heads Perpendicular upward, a silken String going round the Middle to support them.

It may be observed, that should the Caterpillars of the branched Kind fasten themselves, to change with a Thread round the Middle like the smooth Class, they could never get their Skin off them, being interrupted by the Silk Thread; therefore Nature, to avoid that Inconvenience, directs them to hang themselves Perpendicular by the Tail, that they may be free from every Thing which might obstruct them in their Time of Transformation.

But perhaps nothing in Nature more deserves our Consideration and Inspection, than the various Methods those of the Moth Kind take to hide and secure themselves from Danger, while in that helpless and inactive State; some bury and change in the Earth about one Finger deep, within a tender Web; others form a strong Case in the Earth, wherein they change to their Chrysalis; some spin a Case of Silk very strong, most of which are nearly in the Form of an Egg; of this Sort there be great Variety, differing in Form, Texture, and Colour, some are long, and small at each End, others are flat at the Ends, some very soft, others so hard as not easily cut with a Knife, some change in Cases within the Bodies of Trees, others folded and spun up in Weeds, float about in Ponds on the Surface of the Water, some in Stalks of Plants without any Spinning, except that which covers the Hole, which they make for their Passage out; some spin up in a fine transparent Web like Gause, others again are composed of so few Threads as scarcely to contain the Chrysalis; the Caterpillar of the hairy Sort generally mix their Hair with their Webs.

The Chrysalids of the Butterfly Kind, many of them appear all over beautifully gilt with burnish'd Gold, others variously spotted with Gold and Silver; some White spotted with Black, others Green striped and spotted with Brown; most of these differing both in Form and Colour.

Those of the Moth Kind, especially of the larger Sort, are of a dirty Brown, and not greatly differing in Form, those indeed of the smaller Kind vary in Form and Colour, but there are none that may be reckoned extraordinary either for Beauty of Colour or Shape, nor has one of them the least Appearance of Mettle on them, as those of the

Butterfly Kind, that being peculiar to them alone, and of them to one particular Class, which are those of the spiked or branched Kind.

The Eggs both of Moths and Butterflies vary so much, that I could not, without many Exceptions, make any Distinction between one and the other.

I shall therefore conclude this Part of the Introduction with observing, that Butterflies never pass thro' the Winter in the Egg State; and on the other Hand, Moths very rarely pass that Season in their Fly State; an Observation that I believe hitherto has escaped general Notice.

There are several Sorts of Nets made Use of to catch Insects, to wit, the Batfolder, the Racket, and the Scithers Net: The Batfolder is made of Musketta Gause, and is form'd like the Batfolding Net made Use of the catch Birds; these may be had at the Fishing-Tackle Shops, by asking for them; they call them Butterfly Traps.

The Method of using the Batfolding Net is thus: On seeing the Insect come flying toward you, you must endeavour to meet it, or lay yourself in its Way, so that it may come rather to the right Side of you, as if you intended to let it pass; then having the Net in your Hands, incline it down to your right Side, turning yourself a little about to the Right, ready for the Stroke; not unlike the Attitude in which a Batman in the Game of Cricket stands, when he is ready to strike the Ball, only his Bat is lifted up, but your Nets must incline rather downward: When the Fly is within your Reach, strike at it forcibly, receiving the Fly in the Middle of your Net, as it were between the two Sockets of the Benders, that being the Part of the Net which best receives the Insect; and not only so, but should the Fly strike against the Belly or wider Part of the Net, the Course of Air caused by the Motion of the Nets, would carry the Fly with it out of the Net between your Hands, which I have often experienced. The Motion of your Hands in catching, must be from your right Hip to your left Shoulder, not at all retarding the Motion, 'till 'tis as it were spent, closing the Nets in the Motion.

You are likewise to remember never to give the Stroke-over-handed, unless the Situation of the Place oblige you to it. Having closed the Net with the Insect in it, immediately grasp both the Sticks in your left Hand, and with your Right lay hold of the bottom Part of your Net, pulling the Gause pretty tight, giving that also to the Gripe of the left Hand, this confines your Fly from struggling. Put then your Hand against the Fly on one Side, and bringing the Top of your Fore-finger on his Body, and with your Thumb on the other, squeeze him gently, then lay your Nets on the Ground, and take out your Fly by a Horn or a Leg, and holding him in a advantageous Manner by the Body in your left Hand, run a Pin thro' the thick Part of the Body, or Chest, perpendicularly, and put it in your Box.

When you pursue a Fly you must catch him when in your Reach, in the same Manner, except its Course is along a Ditch, on the Left-hand Side of you, and then you will not be able to touch it, the Position being very aukward; in this Case you must overtake it, and turning nimbly about, the Position will then be as in the first Case; the Fly then coming to the right Side of you. I having given you sufficient Instructions for the Use of the Batfolder, I shall next proceed to the Racket Nets.

Which are form'd of Wire about the Size of a Raven's Quill, turned round to a Circle, bending the Ends outwards by way Shanks, which are made fast in a Brass Socket; this Circle or Ring of Wire is covered with Gause, and bound round with Ferret; a round Stick of about two Feet in Length is fitted to this Socket, by Way of Handle. These Sort of Nets are what an AURELIAN should at all Times carry about him; a Pair of these of about six Inches Diameter are the most convenient for that Purpose. The chief Use of these Sort of Netts are for catching Moths, sitting against a Tree, Wall, or Pales; or a Moth or Fly sitting on a Leaf, may be conveniently caught between a Pair of these.

The Scithers Net are no more than a small Pair of these Racket Nets, fixed on two Pieces of Iron which are rivetted across each other, with two of the Ends turn'd round in the Form of Rings, for the Admittance of the Thumb and Finger; in short, a Pair of Toupee Irons, or Curling Tongs, such as is used by a Hair-Dresser, are very well adapted for this Purpose, with a round Net fixed to the End of each Tong with binding Wire, or small Twine well waxed; these Nets are principally adapted to take small Moths, &c.

I shall in the next and last Place proceed to inform the young AURELIANS, what Apparatus is necessary, when he goes out to the Field in pursuit of Insects, and the Method by which they are extended, or set, and preserved, when brought home, and that he may not be inconveniently loaded with more Things than is necessary; he should consider his principal Pursuit, before he sets out, if his Intention is to spend a Summer's Morning in some delightful Woods, where he may expect to find plenty of Flies, he need take no other Net with him, than the Bat-folding Net, one of the Sticks of which may be used as a Walking-stick, and the other, may be made to take in Half, or put together at Pleasure, by a Brass Socket in the Middle, and carried convenient with the Benders in a Canvas Bag under the Coat; it will be proper to take two large Chip, or Dutch Boxes, cut down pretty flat, and lin'd within-side, Top and Bottom with Cork of about a Quarter of an Inch or somewhat less in Thickness, which should be pasted over with White Paper, besides those, he should be provided with another smaller Box for Caterpillars, in Case such should fall in his Way; in the Lid of which should be cut a Hole, as large as will about admit your Thumb to go in easily; this must be stopt with a Cork close fitted, so that small Caterpillars may not get out; together with these, he must take with him a Pincushion well supply'd with Pins of different Sizes, for the different Sizes of Insects, which may be taken, and to be careful not to stick a small Fly or Moth, with too large a Pin, which will certainly destroy it, by putting the Joints of the Wings out of place, for such Insects as are disjointed, will never set well, and fall to pieces in a short Time.

And whereas some of the smaller Flies or Moths, by being indued with so small a Quantity of Humidity, are very apt to dry in the Box, soon after being stuck with a Pin. The AURELIAN should take with him, a Quantity of such Card Braces, as are described on the Setting-Board in the Twentieth Plate, and expand or set their Wings before he goes out of the Field, or rather as soon as he perceives them to be dead, otherways 'tis impossible to do it afterwards.

It may not be improper to mention some other Conveniences, which I have often found very necessary, such as a pretty large Clasp Knife, and some Needles and Thread. The First he will find useful on many Occasions, and the Second is necessary in mending the Nets, in case he should happen to tare them, and to repair other Disasters, which are incident to People who frequent Places where such sharp Things as Thorns and Briars grow.

Having return'd Home with your Insects, look in your Boxes, and observe which of them is fit to sett, such as is dead, but not stiff, are so; then proceed to manage with them as follow: Take a Fly, and observing if the Pin be perpendicularly run thro' the Body, place it on the Setting-board, then take your Point and gently raise one of the upper Wings, 'till such time as the Tip be even with the Nose of the Fly; this done, fix one of your Card Braces on that Wing, to prevent its giving Way; do the same by the Wings on the other Side, and your Fly will be properly extended: Let the Brace remain on the Wings of Butterflies a Fortnight, on those of large Moths a Month.

The Setting-boards of which it is proper to have three or four, should be venear'd over with Cork, near a quarter of an Inch in Thickness, and cover'd over with white Paper, smoothly fixed on with Gum Arabick. The Point which is made use of to sett the Flies, is nothing more than a common large Needle, fixed in a kind of Handle.

Altho' many and various hath been the Methods tried to preserve the Flies in Cabinets, from small Insects which destroy them, by eating away their Bodies, yet all Attempts have been hitherto fruitless; I therefore think it necessary in this Place, to mention a Method which I think will effectually do the Business.

Take the Drawer before it be lined with Cork, and set it some Distance from the Fire, so as to obtain a little Warmth, then with a small Quantity of *Unguentum Serulium,* or Ointment of Russet, on a Woollen Rag, rub it all over, inside and out, pretty strongly; then do the same by the under Side, and the Cork which you propose to line it with, and in covering the Inside of your Drawer with White Paper; on no Account make Use of Paste, as some Sorts of those Insects which destroy them Flies are very fond of it; but strong Gum Arabick instead.

The Ointment may be had at the Apothecaries, and one Ounce is sufficient for twenty Drawers.

AN EXPLANATION &c.

The COMMA
The HOP. LUPULUS, mas & foemina, C.B.

THE *Caterpillar (a)* of the *Comma Butterfly*, which generally feeds on the Leaves of the Hop, but is sometimes found on the Nettle, is very slow of Motion, and may be taken from the latter End of *July* to the Middle of *August*, about which Time it suspends itself by its Tail to the Branches, or under Part of the Leaves ef the Hop by a Web, which, though very fine, is so strong, that unless great Care be taken in separating them, you will pull the *Caterpillar* asunder; it hangs in this Manner about twenty-four Hours, then changes to the *Chrysalis*, as at *(b) (b)* in which State it remains about fourteen Days, and then produces the *Butterfly*, called *Comma*, from a white Mark on the under Side of the under Wing, resembling that Stop in Printing. *(c)* is the Female, flying to shew the upper Side of her Wings; she is larger, her Colour paler, and her Wings not so much indented as those of the Male, which is represented settled on the Leaf of the Hop at *(d)*, to shew the under Side of her Wings. This Fly hides itself during all the cold Season, and appears again in the Month of *April*, much faded in its Colour, when it lays its Eggs on the young Sprouts of the Hop and Nettle, which are hatched about the Middle of *May*, go through the same Changes as above, and produce a Fly by the latter End of *June*, which lays the Egg that produce the first-mentioned Caterpillar

N.B. The *Comma* is very swift in Flight, and timorous when settled, that it is difficult to get near enough to lay the Net over it; they fly in Lanes, by Bank Sides, often settling in dry clayey Places, and against the Bodies of Trees.

The BURNET MOTH
Common Meadow Grass
Gramen pratense minus vulgatissimum. Dr. Hill.

THE *Caterpillar (e)* feeds on *Hay-Grass*, where it may be found during all the Month of May, and sometimes in *June*; but those which are found so late, are generally stung by the *Ichneumon*. It goes into *Chrysalis* from the Middle of *May* to the Beginning of *June*; in which State it is inclosed in a yellow silken Bag or Web, fastened to the Sides of the Stalks of the Grass, as represented at *(f)*; where it remains about fourteen Days, then produces the Moths *(g)* and *(h)*; the Hen *(g)* is largest, and her Colour more tending to Yellow than the Cock *(h)*, which flyes about in Search of her, while she sits waiting for him in the Leaves of the Grass: They may be easily taken with your Hand when settled; and when taken, will lie for some Time as if dead, though unhurt.

N.B. They remain in the *Caterpillar* State during all the Winter, and in *June* the *Caterpillar*, *Chrysalis*, and *Moth*, may be all found at the same Time. *(i)* is the *Chrysalis* of the *Burnet Moth* taken out of its Web. An *Ichneumon* Fly often lays its Eggs in the Body of the *Burnet Caterpillar*, which are there hatch'd, and produce small white Maggots, which feed on its Flesh. The *Caterpillar*, notwithstanding it is infested with so many of these Devourers, that they swell his Body much beyond its natural Bulk, sometimes lives till the Time they generally begin to spin their Web, in which they inclose themselves when about to go into *Chrysalis*, but is destroy'd by these Insectophagi before he can complete it. When these Maggots are near full fed, they eat their Way out through the Skin of the *Caterpillar*, as at *(m)*, and leave it shrunk and dying, as represented at *(k)*. In a short Time after they go into *Chrysalis (n)*, and in about ten Days more produce Flies like their Parent *(l)*.

The PLUMED MOTH

THE *Caterpillar*, which is so small, that it is difficult to be found, feeds on Grass by the Sides of Ditches. They are easiest to be taken in the Beginning of *June*, when they are going into *Chrysalis*, as at *(o)*; in which State they remain about fourteen Days, then produce the Moths *(q) (r)*, which settle on the Leaves of the Nettle, chiefly in Flocks.

Plate I

The Comma, Polygonia c–album *is the first butterfly in* The Aurelian *to be illustrated by Harris. It is one of only five species which successfully hibernate as butterflies in Britain, a habit which did not escape Harris's notice. Priority is given to hop as its food plant, for this was much more common in years gone by when a brew-house was usually included in farm and cottage out-buildings; London in the mid-eighteenth century was still surprisingly rural. It is now common enough on nettle, although during the last century the numbers of this deceptively ragged-looking butterfly declined until, at the turn of the century, it was almost confined to a few counties bordering Wales. Now, happily, it can be found over most of southern England again. Early in the eighteenth century a pale form of this butterfly was recognized, now known as* hutchinsoni, *which is the one depicted by Harris. The Comma was first described as an English species by Thomas Moffet.*

From time to time when giving his account of the insects in The Aurelian, *Harris makes reference to, and sometimes illustrates chalcid wasps or ichneumon-flies. These 'insectophagi', as he calls them, use many methods of attack on lepidoptera, sometimes laying eggs in or on the eggs, caterpillars or chrysalids, developing within, and emerging in due course as perfect insects from the relics of their hosts. There are many hundreds of species of these parasites, whose life cycle seems to us to be so unsavoury, yet the numbers of some of the more destructive insects are thus kept within bounds; which is very little consolation to the breeder of moths and butterflies! The food of the Six-spot Burnet,* Zygaena filipendulae, *shown on the plate, is bird's-foot trefoil, which is fairly widespread over Britain.*

From the pure white colour of the wings, the little Plume Moth (of which there are several different species) appears to be the common Pterophorus pentadactyla, *the tiny caterpillar of which hibernates to feed on bindweed in the spring.*

To the Right Hon.ble the EARL of BUTE   Groom of the Stole to his Royal Highness

The Prince of Wales, &c. &c.   This Plate is Humbly Inscribed by his most
                                      Obedient & devoted Servant Moses Harris.

M.r Harris ad Vivum f.

To the Right Honble EARL BROOKE &c.
His Lordship's most

This Plate is Humbly Inscribed by
Obedient & Devoted Servant
Moses Harris.

Mo: Harris ad Vivum

Plate II

The PRIVET HAWK
Privet. Ligustrum vulgare. Dr. HILL.

THE *Caterpillars* of the *Privet Hawk*, when first hatched, have the Horn at their Tails very long, in Proportion to their Bodies, as represented at *(a)*. In their second Skins their Shape and Colour is much the same as in their third, in which they are shewn at *(b)*, in a Posture they are frequently found in, when not feeding or seeking their Food: In this Skin they are beautifully frosted with yellow Spots, which disappear in the fourth; but ample Amends is made for this Loss, by the purple Edging of the under Part of the white Streak, which go obliquely across its Sides; and the Feet likewise are adorned with two Lines of fine Purple; with this Skin the Roughness goes entirely off, and it appears in its sixth or last quite smooth, as represented at *(c)*; their common Food is the Privet though they are sometimes found on the *Lilack*, their Colour being so near that of their Food they are not easily found; but if you look carefully on the Ground under the *Privet* or *Lilack*, if there are any Caterpillars on the Tree you will find their Dung, which is of the Shape, Size and Colour as represented at *(d)*; and by it you may be directed to the *Caterpillars* above. This *Caterpillar* is so very tender, that if it falls from the Tree to the Ground, which they seldom do except in rainy Weather, it generally proves fatal to him. But the Great Creator, whose Prudence is manifest in even the smallest of his Works, has armed him sufficiently against Accidents of this Sort, by providing them with a Roughness at the bottom of the middle Feet, and the two Holders at the Tail, resembling the Teeth of a File, with which they hold so fast to the Branches they are on, that if you endeavour to pull them off with your Fingers you will certainly destroy them; the best Way to take them therefore is to cut off the Branch on which they are found with a Pair of Scissars, for a Blow against the Leaves or Branches is capable of destroying them. They are full fed as at *(d)* about the Beginning of *August*, at which Time their Backs are tinged with Brown; they then crawl down the Tree, and go into the Earth about a Finger deep, where they spin a thin weak Web, and in about five Days go into *Chrysalis (e)*, which is remarkable, on Account of a Protuberance like a Nose, which contains the Proboscis of the *Moth*. In this State they lie till the latter End of *May*, or Beginning of *June*, in the Year following, when they produce the *Moths (f)* and *(g)*. The Cock *(f)* is deeper coloured and smaller than the Hen *(g)*, which seldom flies, but when she is in Search of a Place to deposit her Eggs, which she does on the branches of the *Privet* or *Lilack*. It will be proper in this Place to give some Account of the manner of rearing this *Caterpillar* from the Eggs, which are when first laid of a light Green, as represented at *(h)*: When the *Caterpillar* is about to eat his Way out, the Colour of the Egg changes to a dead White; this generally happens in about a Week after they are laid, but the Time is uncertain, for some have laid three Weeks before they come out; as soon as they are hatched lay a Sprig of Privet gently over them, and they will get on it and feed; then put it into a Bottle of Water in a Cage, which must be large enough to hold two Bottles, that they may creep from one to the other when they want fresh Food, which they must be supplied with at least every other Day; for the Juices of the *Privet*, when kept too long in Water, will be so much impoverished or diluted, as to flux the *Caterpillars* and kill them. This Disease is sometimes carried off by the Perspiration being so much increased, as to stand in Drops about the Head, and by this Means they generally recover, and produce *Moths*; but many others, that suffer by this Neglect, will be apparently strong over Night, and found dead next Morning. This Disease is more fatal to the *Caterpillars* of the *Emperor, Eyed Hawk*, and *Poplar Hawk*, than the *Privet*; no other Caution but this is necessary in breeding them; only it will be proper, when they are near full fed, to lay a sufficient Quantity of Mould in the Bottom of the Cage, for them to go into when about to change into the *Chrysalis* State. They are seldom taken in the *Moth* State; but when they are, you must be careful how you handle them, for the middle Joints of their Legs are armed with sharp Thorns, with which they will prick your Fingers in struggling to get loose.

The TORTOISE-SHELL FLY

THE Eggs of this Fly are laid on the uppermost Part of the Stalks of Nettles in great Numbers in the Beginning of *May*: The *Caterpillars* when first hatched are of a light yellowish Green, and cover the upper Part of the Nettle with a Web, in which they herd all together. When they shift their first Skin, the whole Colony removes to a fresh Place, leaving their Skins hanging on the deserted Web; indeed, their Aversion to the Place where they shift their Skins is very remarkable; for although there is sometimes Plenty of fresh Food near it, they always retire to a considerable Distance. In the third Skin *(i)* they make another Remove, still herding altogether in a Web; and now being quite Black, and greatly encreased in Size, they cover the Tops of the Nettles in such a manner, as to make them appear as if enveloped in a Piece of black Cloth for six or seven Inches downwards; in this Skin they are often mistaken for the *Caterpillar* of the *Peacock*.

In the fourth Skin, being too bulky to live together in such Numbers without being troublesome to each other, they separate into Companies of six or seven together, and getting on the under Side the Leaf, with their Web they draw its Edges almost close together, making it hang down on the Stalk, as if dead; and in this manner they feed in their fourth and fifth Skins, and in their sixth and last they separate entirely, and straying about, devour the Nettles in such a manner, as to leave nothing but the Stalks and Fibres of the Leaves; in this Skin they are not so dark coloured, being more of a yellow Cast, as they are shewn in the Plate at *(k)*. About the Beginning of *June* they are full fed, and then some of them fasten their Tails by a strong white Web to the Sides of the Stalks, and under Sides of the Nettles, and go into Chrysalis; others go to the Covings of Walls, Eaves of Houses, and Pales, near the Places where they fed, and hanging themselves up by the Tails, go into Chrysalis. Now the first mentioned Chrysalides are of the Colour of burnished Gold, exceedingly brilliant, as represented at *(l)* The others are so different, that Persons, not well acquainted with them, would scarce believe they were produced from the same Caterpillars, being of a dirty Brown, sometimes spotted with Gold. The Cause of this extraordinary Difference in their Colour, is perhaps beyond the Reach of human Understanding; but the Reason that appears to me the most probable, is that those Chrysalides, which are found in open Places, are nearly of the same Colour of the Places they hang against, that they may not be easily found by Birds that would destroy them. In this State they remain about a Fortnight, and then produce the Butterflies *(m)* and *(n)*; the Female *(m)* is larger, and her Colour more upon the bright Orange than the Male, which is represented sitting close wing'd on the Privet to shew the under Side of his Wings. These Butterflies when they first come out of the Chrysalis, emit from their Tails a Liquor of the Colour of Blood, as the *Peacock* and *Admirable* also do, which being found on the Boards of Barns and on Walls, probably gave Rise to the Accounts we meet with in Histories of its having rained Blood. From this first Brood another proceeds, the Caterpillars of which may be found of different Sizes, during all the Summer, producing Flies larger and finer coloured than the first Brood. They fly about Places where Nettles grow, settling on the Heads of the Teazle Furzes of the Dock, and Blossoms of the Thistle. There is another Species of the Tortoise-shell, smaller than these, the Caterpillars of which are nearly black; in their last Skin they are found at the same Time, and the Flies, tho' of the same Shape, are not near so beautiful as the above described.

The PURPLE EMPEROR
Common Sallow. Salix. C.B.

Plate III

ON the 26th of *May*, in the Year 1758, Mr. *Drury*, an ingenious *Aurelian*, in searching for Caterpillars, beat four off the Sallow, near *Brentwood*, in *Essex*; which in their Shape and Motion differed from any hitherto discovered; being furnished with two Horns, of the same hard Substance as their Heads, resembling the Telescopes of a Snail, and in their progressive Motion seeming rather to glide along like that Animal, than crawl as most Caterpillars do. Struck with the Oddity of their Appearance, and knowing I was about a Work of this Kind, he was so obliging to give me one of them which I fed on Sallow, and found, that excepting the above mentioned Particularities, it greatly resembled the *Hawk* Tribe, having a Point or Horn in its Tail, its Body being green, beautifully frosted with minute yellow Specks, having likewise seven diagonal Lines of a pale Yellow on its Sides, and when at Rest, generally sitting in the Posture these Caterpillars do. It is shewn in its different States of Motion and Rest at *(aa)* on the Third of *June* it forsook its Food, and wandered about the Cage, as most Caterpillars do when about to change, in Search of a Place where they may be least liable to be molested while in the Chrysalis, by the rest of their Brood, and where they may be most securely concealed from those Animals that would destroy them; on the Fourth, it fastened itself by its Tail to the Backside of a Sallow Leaf, and on the Sixth at Night, changed into a Chrysalis of a beautiful Pea Green, with a Bloom of Pearl Colour on it; and what is more remarkable, the diagonal Lines, which crossed the Sides of the Caterpillar, are seen in this State, though the Colour is fainter, as may be seen at *(b)*. This being the Chrysalis of one of the finest Flies in this Part of the World, Providence seems to have taken peculiar Care for its Preservation in this defenceless and tender State, by making its Colour so like the Leaf it hangs on, that it might escape the Search of a very nice Eye. In examining that Part of the Chrysalis which contains the Wings of the Fly, I was confirmed in my Opinion of its being the *Purple Emperor*, by observing, that the square Points of the under Wings projected beyond the rounded Extremity of the upper Ones; this Conformation of the under Wing being peculiar to that Fly. On the 22nd at Night, a few dark Spots were visible on the Wings, and the next Day I found more on different Parts of the Body, which spread gradually till the whole Fly appeared black through the transparent Chrysalis, and about Eight in the Evening, to my unspeakable Pleasure, it produced the Male *Purple Emperor*. Here I hope I may be indulged in expressing my Gratitude to my generous and worthy Friend Mr. *Drury*, for the Discovery of the Caterpillar of one of the most beautiful Flies in the Universe, and which had hitherto eluded the Search of the most skillful and industrious Aurelians; and for the many other Obligations I am under to that ingenious Gentleman. The Colour of this Fly is changeable, according to the different Lights it is viewed in. For in one it appears of a sooty black, and in another the Eye is suddenly dazzled with a resplendent Glow of fine Purple; so that by frequently turning the Fly into different Positions, the Colours play and shift through all the Gradations, from a sooty black to a bright purple, in such a manner, as to entertain the Eye with a delightful and amazing Variety. I have endeavoured to represent the Male flying at *(c)*, to shew the upper Side of his Wings: and although it is impossible for Painting to come up to the Beauty of the Original, yet I hope I have so far succeeded in my Attempt, as to give at least as good an Idea of it, as any Figure hitherto published can do. The Female is painted settled on the Sallow at *(d)*, to shew the under Side, in which she differs little from the Male; but on the upper Side of her Wings, she falls very short of him, being of a dirty Black, without the least Tinct of Purple. Mr. *Nixon* took a Female, which laid five Eggs the twenty-first of *July*, three of which produced Caterpillars the sixth of *August*. This Gentleman endeavoured to raise them, and tried them with several Sorts of Growths; but the Sallow being omitted, they all perished. From this we may be certain, that they are in the Caterpillar State during the Winter. It is a very difficult matter to catch them during their Flight; for they generally hover like a Kite about very high Oak and Ash Trees: and tho', when they remove from one high Tree to another they skim lower than at other times, they do it with such Rapidity, that the Eye can scarce follow them. They delight to settle on the Oak and Ash, creeping from one Leaf to another to sip the Dew, at which Time they may be easily caught. For this Purpose, you must be provided with a Pole fifteen Foot long, with a Net at its upper End, the Mouth of which, when you have covered the Fly, is drawn together by a String, as a Purse is. These Flies are found in the greatest Plenty at *Comb-wood* near *Kingston* upon *Thames*.

The Small GREEN HOUSE-WIFE

THE Caterpillars, which produce this Moth, may be found by beating the white Thorn from the Middle of *May* to the Beginning of *June*, about which Time they are full fed, and appear as represented in the Plate, at *(e)* and *(f)*. The Caterpillars differ very much in Colour, some appearing of a darker, and some of a lighter green, others of a redish brown; but the green Sort are more frequently met with. After spinning themselves up in a slight Web, not unlike a Net, they change into Chrysalis, as at *(g)*, in which State they lay about twenty Days, and produce the Moths, *(h)* and *(i)*, of which *(i)* is the Cock. They may be taken in the Moth's State, by beating the Hedges of white Thorn, which sets them on the Wing, and, as they fly slowly, may be easily taken in your Net.

The Least ERMINE

THE least Ermine Moth generally begins to lay her Eggs the Middle of July on the Branches of the white Thorn and black Thorn, in which State they continue during the Winter, and are hatched about the Middle of *April*; when the young Caterpillar inclose themselves in a Web, together with some Leaves of their Food. When they have devoured all that which is within their Web, they make a fresh Spinning, still inclosing their intended Food. In this Manner they feed till the Beginning of *June*; and now being grown much larger in Size, the whole Colony begins to expand or spread their Web in several Divisions, destroying the Bushes in such a Manner, that sometimes a whole Hedge of white Thorn will be stript of its Leaves. When they are full fed, as at *(k)*, they retire in separate Companies to the inmost Cells of their Spinnings, where they hang themselves up by the Tail in Clusters, and change into Chrysalis, as at *(l)*. This generally happens about the Beginning of *June*. They remain in Chrysalis about fourteen Days, and then produce the Moths *(m)*, *(m)*. The Cock is represented flying, and is darker than the Hen. They fly by Day, and are very common.

The Purple Emperor, Apatura iris was first described by the great English naturalist, John Ray, in his Historia Insectorum, *published posthumously in 1710. We can imagine Harris's delight in rearing this most superb butterfly from the caterpillar given to him by his friend Dru Drury. Benjamin Wilkes, a fellow Aurelian, also mentioned that this butterfly could be taken in Coomb Wood; this was the location where Harris said it could be found in 'greatest plenty'. Nowhere in England can it now be said to be common, although it is not so rare as the kite to which Harris likened it, hovering above a tall oak tree! It must be remembered that when* The Aurelian *was written, a short walk would bring one from London into the countryside and, as late as a hundred years beyond that, pigeons were falling an easy prey to peregrine falcons in the City. The use of the high net for catching* iris *set the fashion for many years, in fact almost to the present day. One can imagine that Harris's net, with a handle fifteen feet long, could be fairly manoeuvrable after some practice — but in more recent times a pole up to double that length has been used, achieving some success but only by very experienced wielders of the net!*

Often very attractive, but sometimes mysterious names were given to insects by early aurelians. The Common Emerald, Hemithia aestivaria *shown also in this plate, was called the Small Green Housewife, an obscure name as it is not usually associated with dwelling-houses.*

The Least Ermine moth was known by the popular name of Little Ermine in the last century but is now known by the appellation Orchard Ermine, Yponomeuta padella. *This is to distinguish it from several allied moths in the same genus, the scientific name of which is frequently written* Hyponomeuta *in order to ensure the inclusion of the Greek aspirate. The names bestowed by early entomologists, particularly the specific ones, are often misleading if translated literally. Padella is derived from* Prunus padus *which is the bird-cherry, and is not really its food plant at all; it eats mainly hawthorn — as Harris rightly observed. Often the caterpillars, hatched from eggs laid during the summer, overwinter in that state but otherwise the eggs live through the winter, hatching in the spring as described in the account. A great deal of damage to hedgerows may be caused by caterpillars of this pretty little black-spotted moth, which sometimes invade gardens right in the London area. The conspicuous silken webs of the insects denote the presence of a colony of these voracious caterpillars.*

Pl. III.

Mos. Harris fect

To the R.t Hon.ble the EARL of MACCLESFIELD President, and to the rest of the Fellows of the Royal Society.

This Plate is Humbly Inscribed by their most Obliged Obed.t and Faithfull Serv.ts Moses Harris & John Gretton.

NULLIUS IN VERBA

To his Grace Henry Augustus Fitzroy, Duke of Grafton, Earl of Euston &c.
This Plate is most humbly Dedicated, by his Graces most Obed.ᵗ & faithful Servant,
Moses Harris.

Plate IV

The PINKUNDERWING
Ragwort. *Jacobea, Vulgaris.* J.B.H. 1057

THE Caterpillar may be found feeding on the Ragwort near *Kentish-town* about the latter end of *July*; some of which are full fed, as at *(a)*, and go into the Earth, where it is soft enough to admit them for the first ten Days of *August*. But the Nature of the Caterpillar is rather to hide itself, than burrow into the Earth, as in Holes or Cracks in the Earth, under Pieces of Dirt, dead Leaves, &c. where they spin, and change into Chrysalis, which is remarkable for its Smallness, in proportion to the Caterpillar, which seems to promise one much larger.

This Chrysalis, which is represented at *(b)*, is, by the Hardness or Thickness of its Shell, deprived of that Motion in the several Rings or Divisions of its Tail, which is common to most other Chrysalides, and will appear, if you do but breath on them. This convinces me they have the sense of Feeling, even in this State, and that to a great Delicacy. The Caterpillar may be found in great Plenty the latter End of August; but those which are seen so late, are for the greater Part stung by the Ichneumon. Some hundreds of the Caterpillars were taken about this time, all of which perished from that cause, nor did one escape those Devourers: therefore, I would advise those who would breed this Moth, to get the Caterpillars as early as they can, and only those which are in their last Skin; for as the Plant will not keep fresh in Water, it will be very troublesome to supply the small ones, till they are fit for changing. If the Place where you find the Caterpillars be far from your Abode, you will no Doubt be careful to take home with you a great Quantity of the Food, and you may keep it fresh for eight or ten Days, if you follow my Method; which is to lay it under a Water-tub or Cistern, and it will imbibe Moisture from the Dampness of the Place, so as to keep it perhaps beyond your Expectation. This Method may be used with any Tree, Shrub, Plant, &c.

When you supply the Caterpillars with fresh Food, which is to be put into the Cage without a Phial, leave a Quantity of the perished Plant in the Bottom of the Cage; for in that the Caterpillar will hide, and change into Chrysalis, which is shewn at *(b)*, in which State they continue till the latter End of *May*, and then produce the Moths, *(c), (c), (c)*.

They fly slow; and may be taken in Fields, Lanes, Gardens, &c.

The CREAM-SPOTTED TIGER

THE Food of all the Tiger Tribe is for the most Part alike, they being what may be called general Feeders, and will feed on any Plant that is not too bitter or biting to the Taste; tho' I observe that every Caterpillar has it beloved and favourite Food, and will eat no other, if that be near. So the Cream-spotted, tho' it will feed on Nettles, Cabbage, Lettice, Grass, &c. yet that which appears most natural to them is Chickweed. The proper Time to seek the Caterpillar is about the Middle of *April*, on old Banks, which face the rising Sun, at which Time and Places they may be found in their last Skin. When they are full fed as at *(d)*, they creep into Holes or Cracks in the Banks, where they spin a weak Web as at *(f)*, within which they change into Chrysalis, which is shewn at *(e)*, as taken out of its Spinning. The Moth appears about the twenty-fourth of *May*, after lying in Chrysalis about twenty-eight Days. The Hen-Moth is represented flying, as at *(g), (g)*: For after she has been cocked, she takes Wing, in search of a convenient Place where she may safely deposit her Eggs, which she lays on the fore-mentioned Places, and seldom is seen flying but on that Occasion. The Caterpillars feed all the remaining Part of the Summer; and, at the Approach of Winter, hide themselves in Holes of the Banks, and remain in a sleeping State till March, when they come forth from their Retreats, and feed again.

The Caterpillar, when full fed, is about two Inches long, and thick in Proportion. It is very black, and covered with brown Hair; the Head and Legs are red as blood, and the breathing Holes on its Side are of a clear white.

The Moth may be taken flying in the Day-time in and about Woods.

The EYED HAWK

Willow. *Salix.*

THE Caterpillars of this Moth were hatched from an Egg of the Shape, Colour, and Size represented at *(a)*, which are generally laid on the Bark of the Willow or Ozier, sticking thereto by a gummy or glutinous Substance, emitted with the Eggs from the Body of the Moth; which is of such a Nature as not to be dissolved by Water. She does not place her Eggs in any particular Order; and the young Caterpillars when hatched immediately seek their Food.

They are to be found most plentiful on such Willows as grow on Ditch-sides, and particularly on those Branches which proceed from the Stumps of such Willows as have been cut down. In these and such-like Places they may be found of different Sizes, from the Middle of *July* to the latter End of *August*. It is easy to discover the Tree on which there is a Brood: For on such Trees the Leaves are eaten in the Manner as expressed in the Plate as at *(b), (b), (b)*; and where such Signs as these appear, you may be certain there are Caterpillars, if not too late in the Season. When you have found them, be careful how you take them from the Branches, to which they adhere very strongly. The Caterpillars, when full fed, as at *(c)*, is near four Inches long, of a fine green Colour, and all over powdered or freckled with minute white Specks. The smaller Caterpillars are rather lighter, as may be seen at *(d)*, which is represented in its third skin, and *(e)*, where they are shewn as they appear when just coming from the Egg.

The most early go into the Earth near the Root of the Willow about the Middle of *July*, and change into Chrysalis, which is shewn at *(f)*; in which State they continue during the Winter, and the Moth appears the latter End of *May*.

The Hen is painted flying at *(g)*, to shew the upper Side of her Wings, and at *(h)*, in a Position which discovers the under Side: *(i)* represents the Cock settled on the Twigs of the Willow.

They fly very swiftly by Night, and are seldom taken.

A small brown Beetle often lays her Eggs on the Caterpillar of the Eyed Hawk, where they are hatched, and the young ones eat their Way into the Body of the Caterpillar, which goes into the Earth, and changes into Chrysalis. Notwithstanding those internal Destroyers, the Moth comes forth from the Shell at its usual Time; but the Caterpillars of the Beetle being now increased in Size, prey so much on the Inside of its Body, as to kill the Moth, on which they still feed, regularly shifting their Skins, till they be full fed, as at *(k)*, at which Time they are about three Eighths of an Inch long, having only six Legs, which are situated near the Head, and their Backs covered with very long Hair. Having eat their Passage out of the Moth, they change into the Nympha shewn at *(l)*; in which State they remain about eighteen Days, and then produce the Beetle *(m)*.

The Caterpillar of the above Beetle is the Insect which makes such Destruction in Cabinets of Flies; for it very frequently happens that, among a Number of Moths, as this Beetle does not confine itself to the Eyed Hawk alone, some one of them will be attended with this Misfortune, and contain within its Body four or five of these Caterpillars, which will sometimes feed for many Weeks after the Moth is dead, eating quite through the Body in many Parts, and running about within Side, raise a very light Dust with their Hairs, which will not only cover the Body of the Moth, but, flying about the Drawer, settle, in more or less Quantities, on the rest of the Flies: Nor is this the only bad Consequence which may be feared from them; for, if the Body of that Moth be not sufficient to supply them with Food till they be fit for their Change, they will betake them to other Insects perhaps of more Consequence.

To destroy the above Caterpillar, and other small Insects which destroy the Flies, many Aurelians put Camphire, tied up in small Mechlin Bags, into their Drawers; yet I would not advise this Method, as it is injurious to the Colours of the Insects, and I know, by many Instances, it hath not the desired Effect.

To take the Dust from off the Wings of the Flies I make use of a Camel's Hair Pencil, which is first made very wet with clean Water, stroaking it gently over the Fly. This will effectually take off all the Dust without wetting the Fly or otherwise doing it the least Damage.

The WHITE SATTIN

THE Female generally lays her Eggs about the Middle of July, which are of a light green Colour, on the Bark of the Willow or Poplar, which she covers over with a white Substance much resembling Silk, and in about twenty Days the Caterpillars appear. They feed the remaining Part of the Summer, during which Time they shift about three Skins, and lie concealed during the Winter in a sleeping State. They may be found full fed, as at *(n)*, about the latter End of June, at which Time they make a Spinning in the Leaves of their Food or in Holes of the Bark of the Tree, wherein they change to a hairy Chrysalis, which is shewn as taken out of the Web at *(o)*, and at the Expiration of twenty Days the Moth appears. The Hen is shewn flying at *(p)* and the Cock sitting close-winged on a Leaf of the Willow at *(q)*.

They may be taken in great Numbers by beating the Bows of the Trees they feed on; and, as they fly very slow, may easily be taken in your Nett.

Plate V

All the species of Hawk-moths are strong fast fliers, hence their name, and the Eyed Hawk-moth, Smerinthus ocellata *is one of the most attractive. Although not the commonest, it is quite frequently found throughout England and Wales. Several of the British Hawk-moths have no 'tongue', the 'Eyed' included, and it is useless to hope for its appearance in the flower garden as it has no power to feed.*

Harris's account of the ravages of the beetle, which is so destructive to collections, appears to be the first published in English; however, it is unclear which species is referred to, as a number have the same pernicious habits. Probably, from his description, it was one of the genus Attagenus, *but the bane of curators today seems to be those species in* Anthrenus — *which includes the notorious* museorum.

The White Satin Moth, Leucoma salicis *is one of those few moths which flourish in the London area, where it is abundant — as it was in Harris's time when he said it occurred 'in great numbers'. It is fairly widespread in other parts of England and Wales but never in large numbers, except locally in Essex and Kent. The English name of the insect is derived from the satin-like appearance of the upper wings.*

PL. VI.

Mr Harris ad Vivum.

To her Grace the *Dutchess of Grafton* *This plate is most humbly Inscribed* *by her Graces most Obedient & faithful Serv.* *Moses Harris.*

Plate VI

The ADMIRABLE

The Great Stinging-Nettle. *Urtica major vulgaris.* J.B.

THE Female Admirable is generally seen to lay her Eggs, about the latter End of June, on the Great Stinging-Nettle; in doing which she flies from one Nettle to another, disposing of her Eggs singly one on a Leaf, and at such a Distance from each other, that sometimes her Store of Eggs will be extended or distributed over two or three Fields. This she does for the more certain Security of some of them; and so careful is she for the Safety of her young Brood, that I have often perceived her, when about to lay an Egg, creep in among the Nettles; which I imagine is not only to place the Egg from the Heat of the Sun, but likewise to see if those Nettles are frequented by Ants, these Creatures being very destructive to Caterpillars.

As soon as an Egg is hatched the young Caterpillar begins to make him a Place of Security, being of a very tender Nature; and to keep himself from the Injuries of the Weather and the Ichneumon which he seems to live in continual Fear of, he incloses himself in a Leaf of the Nettle, by drawing with his fine silken Threads the Edges of the Leaf close together: Here when inclosed he feeds, eating away that Part of the Leaf which is next the Stalk, which causes it to hang down as represented at (a): When he hath destroyed as much of the Leaf as renders it no longer a Place of Safety, and after shifting his Skin, he forsakes his ruined Habitation to go in Search of a Place proper for a new one, which he makes like that already described. In this Manner he proceeds till such Time as he is grown so large that one Leaf is not sufficient to contain him, when he creeps up towards the Top of the Nettle, where he spins himself up within the Leaves after eating the Stalk almost through, which makes it hang down in the Manner expressed at (b). Sometimes he may be found in another Sort of Spinning, which he makes by drawing together the Tops of two or more Nettles that lay most contiguous to him; and it often happens, as some one Nettle, whose Assistance he wants to compose his House, may be at too great a Distance; to make it incline more ready for his Purpose, he eateth away the Stalk on the further Side, causing it to bend towards the intended Place.

In searching for this Caterpillar you will frequently percieve a Nettle hanging down as if it were broke, with its Leaves quite withered and dry. This is done by the Caterpillar of the Admirable; and if you search among those dead Leaves you will find his Spinning, and most likely the Caterpillar or Chrysalis, for he seldom does this but when ready for his Transformation, which happens with the most early Caterpillars about the Beginning of August.

The Caterpillars are of various Colours, some appearing of a light yellow, or Amber-Colour, as at (c); others almost black, like that at (d); both which are seen in their last Skins. The younger Caterpillars are black, freckled with small Specks of Yellow; as at (f).

When they are full fed, they generally fasten themselves up by the Tail within their Spinnings, and change to the Chrysalis, though they sometimes may be found in that State hanging openly under a Leaf, as at (e), or any other Place they find convenient. Why they change, thus naked and exposed, contrary to their Nature of concealing themselves as well in this State as that of the Caterpillar, is what, with any certainty, cannot be accounted for: but the Reason that appears most likely to me is, that the Earwigs, which often, in great Numbers, get into their Inclosures, obliges them to retire, and, being near the Time of their Transformation, are too weak to make a fresh Spinning; for I have often found, when in Search of the Caterpillar, their Spinnings crowded with those Vermin.

The Chrysalis is of a Pearl-Colour, covered with a fine Bloom resembling that which is seen on the Plumb; but the Chrysalides which have this Appearance are those only which are found in the Fields; for those produced from Caterpillars which are fed within Doors, are of a dirty Brown, sometimes embellished with small Spots of Gold. They lay in Chrysalis twenty-one Days, and then produce the Fly called ADMIRABLE from the great Variety and Beauty of its Colours. The Male is seen flying at (g), shewing the upper Side of his Wings. The Female is represented at (h), sitting on the Nettle. She is larger than the Male, and may be known by an additional white Spot which is situated in the red Part of the upper Wing.

They fly in unfrequented Lanes and Places which are over-run with Brambles, on the Blossoms of which they frequently settle. They are very quick sighted, extremely timorous and swift of Wing, and, if they find they are pursued, will betake them to some high Tree where they will stay till their Fright is over, or till such Time as they think the Danger past.

The SMALL MAGPIE

THE Caterpillars of the small Magpie are produced from Eggs of a yellow or buff Colour, which are laid on the Great Stinging Nettle by the Hen Moth about the Beginning of June. The young Caterpillars, as soon as they are hatched, divide or separate themselves, each one infolding himself within a Leaf, the Edges of which is fastened together by a white Spinning, much after the Manner of the Admirable at (a). Within such Spinnings they continue to feed till the Approach of the cold Weather, when they begin to change their Colour, which was of a light transparent Green, to a red, or, rather, Rose-Colour, as is shewn at (m). They then forsake their Food, and each Caterpillar spins himself up within a thin buff-coloured Case, through which the Caterpillar may be plainly seen as described at (n). When they have laid in this Manner about one Month, or more, they lose their Redness, and appear of the same Colour with the Cases which contains them. They then remain without any farther Alteration till the Middle of May, about which Time they change to brown shining Chrysalides, as represented at (o); and in about fourteen Days the Moths appear. The Hen is seen flying at (p), shewing the upper Side of her Wings.

They seldom fly far from the Place where they are produced, for, though on disturbing the Nettles they fly out in great Numbers, yet they immediately take to the Nettles again, settling on the under Side of the Leaves with their Wings spread flat.

THE ELEPHANT

White Ladies Bedstraw. *Gallium Album.* Ger.967.

Plate VII

THE Caterpillar of the Elephant, or Bedstraw-Hawk, is produced by Eggs of a greenish white Colour, which are laid by the Hen Moth, on the White Ladies Bedstraw, in the Manner represented at *(a)*. These Eggs are commonly hatched in about eight or nine Days. When the young Caterpillars come forth from their Eggs they appear of a fine green Colour, as may be seen in the Plate where they are crawling from the Eggs, and at *(b)* and *(c)*, where they are shewn in their second and third Skins. Many continue to be of this Colour till they appear in their fifth Skins, though others lose that Appearance sooner. When they are in their sixth and last Skins, they are of a deep brown on the Back, which softens downwards, towards the Belly, to a buff Colour. It is likewise marked all over with short Strokes of black, which crossing each other compose irregular small Squares. On the fourth and fifth Joints, or Rings, from the Head, is placed two Marks, which have the Appearance of Eyes, and which are incircled with a deep black. From hence to the Head it grows taper, the Head being very small: And it is remarkable in this Caterpillar, that it can draw the three first Joints with the Head into the Body, or extend them, at Pleasure; the Manner of which is shewn at *(d)*. When they are full fed, as at *(e)*, they make a light Spinning among the Food, in which they lay four or five Days, and are greatly diminished in Size before they change to the Chrysalis shewn at *(f)*, which must be consequently small, considering the Size of the Caterpillar by which it is produced. They remain in Chrysalis during the Winter, and the Moths appear the latter End of *May*.

The Hen is seen flying at *(g)*, shewing the upper Side of her Wings, and the Cock at *(h)*, as settled on the Bedstraw in a Position which discovers the under Side. I never knew any of them to be taken in the Fly State.

As the Bedstraw generally grows in or by the Side of the Water, it sometimes happens that the Caterpillar falls in, either through the Weakness of the Plant and the Weight of so large a Caterpillar, or the Wind violently shaking the Food; but he generally survives this Accident, for he immediately rises to the Top of the Water, which bears him up till, by the Currents driving him, or his own strugling, he attains to some Part of the Plant, by which he crawls up and saves himself.

It is to be observed that the aforementioned fourth and fifth Joints are considerably larger than the rest, and appear to swell out. This induced me to think that this Part contained a Quantity of Air which was placed there by Nature on Purpose to assist him in the above Extremeties; and what seemed to confirm me in this was, my being in Search one Day after these Caterpillars, by Accident let one fall into the Water, when I observed the Head with the three smaller Joints adjoining, was above, and the thick Part even with, the Surface.

This Caterpillar being quite naked, and in no Way capable of defending itself, is frequently destroyed by his destructive and fatal Enemy the Ichneumon, which settling on the Back of the Caterpillar layeth her Eggs thereon, together with a strong gummy Matter, which causeth the Egg to adhere so firmly as not by any Means to be taken off without tearing, or otherways wounding, the Caterpillar. When the Egg is ready to hatch, the young Caterpillar within does not proceed from the Egg outwardly or upward, but, making his Way through that Part of the Egg which is fixed to the Back of the Caterpillar, eating downward into its very Body: Nor does the Caterpillar, which is attended with this Misfortune, appear any Ways ailing, but eateth its Food freely; and, when full fed, maketh its Spinning and changeth to the Chrysalis; but, about the Time when the Moth is expected, the Ichneumon appears, which is shewn flying at *(i)*. The Elephant Chrysalis at *(k)* is represented with Part of it broke open, shewing the Ichneumon within as it appears when near the Time of its coming out in the Fly State.

The SMALL WHITE CHINA MARK

THE Caterpillars shewn at *(l)((l)* feeds, spun up in the Duck-Weed, which is often seen to cover entirely the Surface of Ponds, and is discovered by little Lumps or Risings on the Weed, such as is represented at *(n) (n)*. These are the Spinnings of the Caterpillars. The Web with which he joins the Leaves together and infolds himself within, is white, and consists of three or four different Coverings or Cases, and are so very strong that it is difficult to pull them open with the Fingers without destroying the Caterpillar or Chrysalis inclosed. Within these Spinnings they change to the Chrysalis, as at *(m)*, and the Moths appear in about fourteen Days, which are represented at *(o)* and *(p)*. The Hen at *(o)* is larger than the Cock, and her upper Wings appear of a brownish Colour, while those of the Cock are perfectly white. There is two Broods a Year. The Caterpillars of the latter Brood change into Chrysalis the Beginning of *August*, and produce their Moths about the Middle of that Month, laying their Eggs soon after. The Caterpillars proceeding from these Eggs lay all the Winter spun up in the Duck-Weed, change into Chrysalis the Beginning of *May*, and produce their Moths in fourteen Days. They fly in great Plenty over the Surface of Ponds.

And, when GOD had created Male and Female of every Kind, He made Man to reign over the Beasts of the Field, and to contemplate the Wonders of His Works; for the Subsistence of which, GOD said, Behold! I have given you every Herb bearing Seed; to you it shall be for Meat: And to every Beast of the Earth, and to every Fowl of the Air, and to every Thing that creepeth upon the Earth, wherein there is Life, I have given every green Herb for Meat.

GENESIS I. 30.

O GOD! how wonderful are Thy Works! For ever blessed be the Name of the Lord!

The companion of Ulysses, Elpenor, was transformed by Circe into a pig. It was from the snout-like appearance of the caterpillar's head that Linnaeus, the great founder of the scientific binomial system of nomenclature, gave Deilephila elpenor its specific name. However, the appellation of Elephant Hawk-moth, as used by Harris, was the one in general use in his time and has been used ever since. It is more pleasing than one which could have ensued from the Linnean name! Although fairly commonly found over most of the British Isles except Scotland, it is much more frequently met with in the caterpillar state, and Harris admits he had not heard of any taken as a moth. It is one of many which are attracted to light, and modern light-traps have shown how often it occurs. The caterpillar of this fast-flying moth has several food plants, not just the one mentioned by Harris, which he calls the 'White Ladies Bedstraw' but which resembles the plant we know as the marsh bedstraw. The name Bedstraw Hawk, which he gives as an alternative name for the Elephant, is the English name we now use for a quite different species Hyles gallii, a rare immigrant. It is interesting to note that many eighteenth-century aurelians, including Albin and Dutfield, when illustrating this insect, depicted the marsh bedstraw with the same designation as Harris. These authors also commented with Harris on the aquatic nature of the food plant. Yet willow-herbs are now probably the commonest food of the caterpillar, and these are plants which grow well in a very dry habitat, as was demonstrated by the colonization of derelict bomb sites in London. It was often thought that the caterpillars were sexually dimorphic, the 'hen' coloured green and the 'cock' brown; Harris made no reference to the misconception in this issue of The Aurelian, *although in a later unpublished plate, illustrated in Riley's* Some British Moths *in 1944, he quite clearly does subscribe to the error.*

The Small White China Mark is now known as the Small China-mark, Cataclysta lemnata; *the 'White' is omitted because only the male is white, as Harris says. Its common name derives from the porcelain-like markings on the wings. It is a common little moth, but unusual in that it is one of a family in which the larvae feed below the water. However, in* lemnata *this happens only in its first instar, that is, until the first change of larval skin. It also pupates just below the surface of the water.*

To his Excellency Baron Kniphaufsen, — Embaſsador from his Pruſsian Majesty, to the Court of Great Britain, — this Plate is most humbly Inscribed? by his Excellency's most Obliged & faithful Servant Moses Harris. —

Plate VIII

The WATER BETONY MOTH

Water Betony. *Scrophularia Aquatica Major.* C.B. Pin. 235.

THE proper Food of this Caterpillar is Mullein, though they are frequently found in Plenty feeding on the Water Betony. The first Appearance of the Caterpillar is about the latter End of *May*, or Beginning of *June*; from which Time they may be found till the Beginning of *July*; therefore the best Time to search for the Caterpillars is about the Middle of *June*, when the greater Part are in their last Skin: when they appear of the Size of that represented at *(a)* they are of a beautiful clear white, prettily spotted, and streaked with black, with some few Stains of yellow: The Head doth much resemble a human Scull, for which Reason it hath sometimes been called the Caput Mortunum, or Death-Head.

It is remarkable in these Caterpillars, that they commonly eat their Skin after they have thrown it off; and, should they be so far neglected as to want Food, they will fall on and devour one another. When they are full fed, they go into the Earth about a Finger deep, where they make a thick Case of Earth, like that at *(b)* which they spin together with their Web, wherein they change to the Chrysalis, which is of a reddish brown, as is represented at *(c)* they remain, during the Winter in this State, and the Moths appear about the Middle of *April*. It is represented flying at *(d)* and in a setting Posture at *(e)*.

They are very seldom taken in the Moth State by reason they fly by Night: The best Place that I know to take the Caterpillars is on the Water Betony, which grow plentifully against the Bank on the Right-Hand Side the Road leading from *Bagnigge-Wells*, towards *Pancras*.

The PEACOCK FLY

THE Female Peacock layeth her Eggs the latter End of *April*, or the Beginning of *May*, on the Top Part of the Great Stinging-Nettle, placing them generally on the Stalk close under the young budding Leaves, to preserve them from the too violent Heat of the Sun, where they are hatched in a few Days. The young Caterpillars, as soon as they approach from the Eggs, inclose themselves together in a very fine tender Web, drawing at the same Time the Leaves to cover them, as much as they can, that they still may receive the Benefit of their Shade; in this first Skin they are of a greenish white, and appear naked and shining, not unlike Maggots; in the second Skin they appear brown and shining, which Appearance continues till they are in their fourth Skin, at which time they are quite black. After the shifting of each of their different Skins, which they always leave in their old Spinnings, they extend their Web farther, and will sometimes divide themselves into two or three separate Colonies, more especially if the Nettles where the Brood is are thinly scattered.

When they are in their last Skin, as at *(g)* they quite forsake their Web, and feed separate, at which time they are about two inches long, and thick in proportion, they are of a fine deep black, and all over powdered with small white Specks; the head, and six hooked Legs, are black and shining; the eight middle Legs, with the Holders behind, are light red, or Flesh-Colour, on the Back and Sides; they are armed with black branched Spikes, which are about a Quarter of an Inch long; at the Bottom of each of these something appears to shine or glisten, as they move along, like small Sparks, or Diamonds, which causes them to have a beautiful Appearance.

When full fed, they hang themselves up by the Tail, in any convenient Place, in the Manner shewn at *(f)*, and, in about twenty-four Hours, the Skin flips off, and the Chrysalides appear, which is first green and tender; but one Hour is sufficient to harden the Shells against the Injuries it might receive by the Plants being shaken by the Wind, or by their being blown against the Place, on or near which they have hung themselves. The Chrysalis is represented at *(h)*, in which State it continues about nineteen Days, and the Fly comes forth. The Female is seen flying at *(i)*; she is much larger, and her Colours lighter than the Male, which is shewn sitting on the Water Betony, with his Wings shut to shew the under-side of them.

They continue in the Fly State during the Winter, and appear very soon in the *Spring*: I have seen them flying in the Month of *February*, when the Snow has been on the Ground.

They fly in Lanes about Banks and Ditches, where Nettles grow, settling on the Heads of the Teasel and Furzes of the Docks.

Although the Caterpillars of the Peacock appear to be sufficiently armed by Nature against the Attacks of the Ichneumon, yet the smaller sort will get between the Spikes of the Caterpillar, where they effect its Destruction, by piercing him with their Tube, through which they convey their Eggs into its Body: And about the Time when the Caterpillar is full fed, and near the Time of Change, the Maggots of the Ichneumon make their Way out on all Sides; leaving the Caterpillar expiring. These Maggots in a short Time spin themselves up in thin silken Bags or Cases; and produce their Ichneumons in about fourteen Days. But provided the Caterpillar is in its last Skin, when pierced by the Ichneumon, though I know that two Days is a sufficient Time for Maggots to be formed, yet they will not have Time sufficient to render him too weak for changing into Chrysalis; so that the Maggots feed on the Inside of the Chrysalis and there change to the Nympha; and about the Time when the Fly is expected, the Ichneumon Flies appear from a small round Hole in the Chrysalis, which they make to facilitate their Escape.

The BRINDLED BEAUTY

The Morel Cherry. *Cerasus*

Plate IX

THE Caterpillars feed on most Fruit-Trees, as well as on Privit Lime, Elm, &c. in the Cracks and Holes of the Barks; of which, about the later End of *April*, the Female Moths deposite their Eggs. The Caterpillars are nearly black in their first Skins, as shewn at *(a)*, where they are represented as coming from the Eggs, but grow rather lighter as they increase in Size; when they are in their last Skin, some appear smaller, and darker coloured than others, which I take to be those which produce the Cock-Moth, and are represented at *(b)*; these are of a deep Cinnamon-Colour, prettily diamonded on the Back with black, and spotted with yellow. The larger and lighter Sort are shewn in a Position which they are commonly seen in when at rest at *(c)*; they go into the Earth about the beginning of *July*, and change into Chrysalis, which is seen at *(d)*, and continue in that State during the Winter. The Moths appear the Beginning of *April*. The Hen is shewn at *(l)*, hanging or creeping on a Leaf, shewing the under Side of her Wings; she may be known from the Cock by her Horns, which appear like fine Threads, while those of the Cock are comb-like, and very broad.

They may be taken in the Day-Time, sitting against the Barks of the above-mentioned Trees, early in the Spring.

The BLACK VEINED WHITE

THE Female Fly lays her Eggs on the White Thorn, about the later End of *June*; and the young Caterpillars, as soon as hatched from the Eggs, inclose themselves in a slight Web, leaving a Passage to come forth to feed, which they generally do Morning and Evening, retiring within their Web in the Middle of the Day, to avoid the Heat of the Sun: In this Manner they feed the remaining Part of the warm Weather, extending their Web as they increase in Size. At the Approach of Winter, they spin a strong Web on one of the Twigs of the White Thorn, wherein they remain without eating during the Winter, and come forth again early in the Spring, feeding very greedily on the Buds and young tender Leaves. Now is the best Time to take them, they being easily seen on Account of their Size, as they lie on their Web altogether, which they do not forsake till they go in search of a convenient Place for their Change, at which Time they appear as represented in the Plate at *(h)*. In preparing for their Transformation, they fasten their Tail to the Twig by a strong white Web, after which they carry a strong Thread, that is composed of three or four double, over their Back, near the Head; this is likewise made fast to the Twig on each Side: In this Position, they retain the Form of the Caterpillar twenty-four Hours, and the Chrysalis appears, which is of a yellow Colour, beautifully streaked and spotted with black, as at *(g)*; they remain in the Chrysalis State twenty-one Days. The Male is shewn at *(i)* flying in a Position which discovers the under side of his Wings, where the Markings are strongest. The Female is seen at *(k)*, she is rather larger, and her black Veins not quite so strong as those of the Male.

They fly in Meadows near Corn-Fields, and as they do not fly very fast, are easily taken in your Net.

In the London area the Brindled Beauty, Lycia hirtaria *is quite common, as it was undoubtedly when Harris looked for it on the trunks of fruit trees in the weak sunshine of March and April. He would have had little difficulty in catching specimens of the male, 'the Cock with the comb-like Horns' after dark for it is strongly attracted to light. Harris was led astray by the frequently held notion that differently coloured caterpillars denoted different sexes.*

The Black Veined White Butterfly, Aporia crataegi *is exceptionally interesting, because it is a butterfly now extinct in the British Isles. It is still common on the Continent, as it was over most of England and Wales in Harris's lifetime, yet it gradually became more local in distribution, until it finally disappeared from all its usual haunts. It is said that the last indigenous specimens were taken in Kent where it was locally abundant until at least 1907. Several theories have been put forward suggesting the reason for this; perhaps the most likely is that it was due to climatic changes in this country, as the butterfly was on the edge of its distribution range. It is, however, clear that to Harris this rather primitive-looking insect was in his day far from uncommon. It was first described by Merrett in* Pinax Rerum Naturalium Britannicarum *in 1666.*

PL. IX.

To the Right Hon.ble Lord Viscount Charlemont, this Plate is Humbly Inscribed by his Lordships most Obliged & Obedient Serv.t Moses Harris.

To the R.t Hon.ble Lady Henrietta   Alicia Wentworth

This Plate is most humbly Dedicated,   by her Ladyships most obliged & obedient Serv.t

MEA GLORIA FIDES

Moses Harris.

Plate X

The PURPLE HAIR-STREAK
The Common Oak. *Quercus Latifolia.* Park. Theat.

THE Caterpillars feed on the Leaves of the Oak, and are found plentiful on such Trees as grow by the Sides of Woods, separate or apart from the rest; they are full-fed about the latter End of May, and are taken by beating the Boughs with a long Pole, provided for that Purpose, while a Sheet is spread under the Tree to receive the Caterpillars which fall. When full-fed, as at *(a)*, they are about the Size and Form of the Millipedes, or Wood-louse of a brown Colour, inclining to the orange, with oblique Markings on the Side; the Head is very small, and lies a little concealed under the first Joint, or Ring; so that, at first Sight, it is not easy to discover the Head from the Tail. They seldom creep, but when they do, their Pace is very slow. They prepare for their Change, by hiding or concealing themselves in the Hollows or Foldings of the Leaves, or against the under Part of the Twigs, as at *(b)* and *(c)*; where they fasten themselves by the Tail, and round the Middle with a very slight and tender Spinning, which is so very fine, as scarcely to be seen by the naked Eye. The Fly appears in twenty-one Days.

The Male is seen flying at *(d)*, shewing the upper Side of his Wings, which are of a dark brown or black, with a blue Spot on each of the upper Wings, which appear of a Brightness equal to Brimstone on Fire. The Female shows her upper Side at *(e)*, which appears of a dark brown; but, on turning it about, or altering its Position, the Eye is agreeably surprized with a fine glowing Purple. The under Side is shewn at *(g)*, which of Male and Female are much alike. They fly high, delighting to settle on the Leaves of the Oak, and are commonly taken in Plenty in *Oak-of-Honour* Wood, near *Peckham*, in *Surry*. The Caterpillar *(f)* is represented as one that has been pierced by the Ichneumon, and the Maggots making their Way through the Skin of the Caterpillar, after which they spin themselves up in white Cases, as at *(h)*, where they are represented as coming to the Fly State.

The SILVER LINES

THE Caterpillars are taken by beating the Boughs of the Oak the latter End of *September* and the Beginning of *October*, which is about the Time they are in their last Skin; they then appear at *(i)* of a light green, marked all over with irregular Spots of white. On each Side is a Line of light yellow, which reaches from the Head to the Tail: A Circle of the same appears round the Neck and the two Legs, or Holders, at the Tail, which are of a remarkable Shape and Length; on each of which appears a Spot of bright red, or crimson Colour. This Caterpillar adheres so very strongly to the Twigs and Branches, that it is very difficult to take them off, if at all, without Damage.

When full fed, they spin a strong Case, which appears not unlike the Bottom of a Boat, wherein they change to a flesh-colour'd Chrysalis, marked or shaded on the Back with Purple, as is shewn at *(k)*. The Moth appears the latter End of *May*. The Cock is shewn flying at *(l)*, and the Hen in a sitting position at *(m)*. There is no very visible Difference; the Cock appears a little less, and somewhat darker than the Hen.

The upper Wings are of a light yellow green, with three white Lines, which cross each Wing in an oblique Direction, which appears to shine a little like Pearl: The Body and Under-wings are of a greenish white, as is all the under Side: The Horns and Legs are red. They seldom are seen flying, but will frequently fall in the Sheets when you are beating the Oak for the Caterpillars, which feed thereon in the Spring. They will lie for some Time in the Sheet as if dead; nor can you tell whether they are dead or living; but you are soon relieved from your Uncertainty, by their starting suddenly, and flying away.

The PEA-GREEN

THE Caterpillar is of a fine Green, and feeds spun up in the Leaves of the Oak. It may be taken full fed by beating about the latter End of May, when it appears of the Size and Form of that represented at *(y)*, which is shewn hanging by his Thread or Web, by which it frequently lets itself down when in Danger or disturbed in its Spinnings. It changes into the Chrysalis *(z)*, infolded in the Leaves about the Beginning of June, and the Moth appears the latter End of that Month. It is seen flying at *(2)*, and in a settling Position at *(1)*. They fly by Day and are very common.

The DUN-BAR

THE Caterpillar feeds on the Oak, and is found full fed the latter End of May. It then appears as at *(n)* of a light transparent Green, with a Mark of light Yellow on each Side. On the two first Rings, or Joints, next the Head, is a Row of small black Specks, which lie all over the Back, reaching from Side to Side; and, on viewing it closely, a few fine Hairs may be discovered on every Ring or Division, it makes a Spinning in the Leaves of the Oak, and changes to a red Chrysalis covered with a fine Bloom, as at *(o)*. The Moths appear in July. It is represented flying at *(r)* shewing the upper Side. The under Side is seen at *(q)*. Some of this Kind of Moths differ so much from each other, that they do not appear to be of the same Species.

The SCALLOPED WINGED BROAD BAR

THE Caterpillar feeds on the Oak and Hazel, and is found about the Beginning of October. It is about an Inch and a Half long, of a brown Colour with a small double Line along the Sides reaching from the Head to the Tail. It has two Protuberances under the Belly, which are situated near the four Holders behind. It goes into the Ground the Middle of October, where it makes a strong Case of Earth, as represented at *(t)*, which is spun and interwoven with their Web. They change into a red Chrysalis, which is shewn at *(u)* as taken out of the Case, the Moth appears the Beginning of May. The upper Side is seen at *(w)*, and the under at *(x)*, which is represented as hanging or creeping on a Leaf. They are seldom taken in the Fly State.

The PAINTED LADY
The Thistle

THESE Flies are not very common; the Reason of which is, all Weathers do not agree with them; yet there are particular Seasons when they are very plentiful, which happens once in about ten or twelve Years. They are then often seen in Town flying in the Streets. The Female Fly lays her Eggs on the Dock and Thistles about the middle of June, depositing them on a Leaf in the same careful Manner as the Admirable. There are various Colours of the Caterpillars, some appearing Dark or nearly Black as at *(a)*, others Lighter and more of a yellowish Cast, as seen at *(c)*. They are found covered with a thin Spinning on the upper Side of the Leaf, in the Manner of that at *(a)*, to secure itself from the Weather and other Accidents, within this Web it feeds, leaving the thin membranous Part to support it in its Habitation; so that the Leaf appears to be eaten but Half Way through. It forsakes its Web when fit for its Transformation, which happens about the Middle of July, and finding a convenient Place in the Shade, fastens itself up by the Tail with a small but very strong Web, and changes into Chrysalis, in which State the Male and Female may be easily distinguished from each other. The Male is of a Dark Brown, embellished with Gold, as at *(d)*. The Female is rather lighter, and ornamented with Silver, as shewn at *(b)*. The Fly appears in fourteen Days. The Female is seen flying at *(f)*, shewing the upper Side of her Wings: The Male at *(e)* in a Position which discovers the under Side. They are taken in the Fly State all the Month of August, and haunt Waste Grounds, Rubbish Hills, &c. settling on the Furzes of the Docks and Blossoms of the Thistle.

The MARMORESS, or MARBLED WHITE

THE Eggs by which these Caterpillars are produced, are, when first laid, of a yellowish Colour; then, in two or three Minutes, changes to a clear and pleasant White. They appear perfectly round and very smooth, which causes them to roll very freely and nimbly about like small Globules of live Mercury or Quicksilver, so that, as soon as voided from the Fly they fall among the Grass to the Ground, rebounding from all Obstruction, and resting at length in some small Crannie of the Earth, their globular or sphere-like Form defending them from the Injuries they might otherways sustain by what they strike against in their Fall; these Eggs are hatched in about fourteen Days.

The Caterpillar feeds on Hay-Grass, thriving but slowly the remaining Part of the Summer, and during the Winter lies concealed, abstaining from food till April, then feeding again on the fresh and tender Grass, becomes full fed about the beginning of June, as it appears at *(g)* they Change into Chrysalis on the ground, as at *(h)* and the Fly appears at the Expiration of Twenty One days. The Male is represented flying at *(k)* displaying the upperside of his Wings, the underside is shewn at *(l)*. The Female is not unlike the Male on the upperside, but the underside appears of a Yellow Cast; in particular the underwings; the markings of which are of an Orange, or Yellow Brown Colour, as may be seen at *(i)*. They fly in meadows, and Fields of Mowing-Grass. They do not range or stray from the Fields in which they were bred, for its common to see them, in a plentiful season, Forty, Fifty, or an Hundred on the wing together, and pehaps in the Field or Meadow adjoining not one to be seen; they may be taken good for about Ten Days, or perhaps a Fortnight; after which they may be seen in Copulation with the Females; which I observe in all the species both of Moths and Butterflies that have come under my observation to appear later then the Males, and tis common, in Butterflies especially, to take the Females in good condition, when the Males are quite wasted and decay'd, they seldom fly high unless when wantonly playing together, when I have seen them fly so high, that my sight has wanted strength to follow them any further.

Plate XI

The Painted Lady, Cynthia cardui *is an extremely strong-flying butterfly, which occurs with little variation in form over most of the world, sometimes migrating in immense swarms and occasionally invading this country in large numbers, as in 1980. Harris did not realize that it was an immigrant and was unaware that it cannot overwinter in Britain in any form, whether as an egg, caterpillar, chrysalis or butterfly. First described by Moffet, the Painted Lady, known in France as 'La Belle Dame', is sometimes called also the Thistle Butterfly because of its partiality for thistle heads, but fortunately its more attractive name has general acceptance. Harris decided in this plate to add, in his enthusiasm, some everyday eighteenth-century debris including shells, broken china and pieces of broken clay pipe, possibly to illustrate the habitat of this butterfly which 'haunts Waste Grounds, Rubbish Hills, &c.'*

The Marbled White Butterfly, Melanargia galathea, *first described in England by Merrett, was known to the early aurelians by a variety of names. To Petiver it was 'the Half-Mourner' but in time this was replaced by Harris's name, The Marmoress (sometimes written as 'Marmoris' or 'Marmores') derived from the Latin* marmor *meaning marble. Confined to the south of England and Wales, this is an extremely local insect, restricting itself, as Harris so acutely observed, sometimes to a single small area in which it may be quite numerous. He was indeed fortunate if such a colony existed close to London, as it may well have done, in view of his very full account of the butterfly. The habit of scattering the eggs at random over grass and rough ground, not usual among British lepidoptera, did not escape Harris's notice.*

To the Right Honourable *Countefs of Dalkeith* This Plate is most humbly Dedicated *by her Ladyships* most Obliged & Obedient Serv.t

AMO

Moses Harris

Plate XII

Abraxas grossulariata *was known for many years as the Gooseberry Moth, from the food plant of the caterpillar, hence the specific name bestowed by Linnaeus. However, both this and Harris's popular name, the Large Magpie, are now superseded by the common name, the Magpie. The caterpillar can sometimes be a pest in gardens where currants and gooseberries are grown, although the latter food plant is not mentioned by Harris — perhaps they were not widely cultivated near the metropolis in his day.*

This account of the Camberwell Beauty, Nymphalis antiopa *is a milestone in British entomological history, for it is the first to be published there of this splendid butterfly. However, it is not really a British insect, although almost every year a few of this handsome species are found there, and in some years they may be quite frequently sighted over a large area. They are of course migrants from Europe, and the occasional invasion, such as occurred in the 'antiopa years' of 1872 and more recently, in 1976, (when it was even more widespread than in its earlier 'annus mirabilis') earned for them the name adopted by Harris, the 'Grand Surprise'. His view that it is a hibernating insect is correct, but in Britain it is unable, except very rarely indeed, to survive the unsuitable climate of the winters. As the Camberwell Beauty is not known to breed there, Moses Harris's worthy intention to make a strict search for the caterpillar and chrysalis must have led to great disappointment!*

The LARGE MAGPIE

The Currant-Tree. Ribes major fructu rubro, Hort. Eyst

THE Caterpillar of this Moth is produced by a small round Yellow Egg which is laid by the Hen on the Branches of the Whitethorn, Blackthorn, and Currant Tree, about the middle of July. They feed the remaining part of the Summer and remain during the Winter on the Branches; adhereing very strongly thereto by their Feet or hind Holders; in this manner, fixed and motionless, I have often found them in the middle of Winter. In about April they begin to feed again and arrive at Maturity as at (f) the beginning of June: they then spin a very weak Web against the Branches wherein they change to the Chrysalis as appears at (g) and the Moth appears in twenty-one days. The Hen is shewn flying at (h). The Cock in a sitting position with his Wings closed at (i). They Fly in the evening after Sunset.

The GRAND SURPRIZE. Or CAMBERWELL BEAUTY

THIS is one of the scarcest Flies of any known in England, nor do we know of above three or four that were ever found here, the first two were taken about the middle of August 1748, in Cool Arbour Lane near Camberwell; the last in St. George's Fields, near Newington Butts, the beginning of that Month; but as these appeared very much faded and otherways abused, I conclude they appear from the Chrysalis, with the Peacock, about the middle of July, and being of that Class tis reasonable to suppose they live thro' the Winter in the Fly state, and lay the Eggs in Spring that produce Flies in the July following; for in the same manner do all the Flies of this Class, and as all that have yet been taken were found flying about Willow-Trees tis the common opinion of Aurelians that their Caterpillars feed thereon; but their Caterpillar and Chrysalis is to us intirely unknown, and the food a mere conjecture. I do intend to make a strict search concerning them and should I make any discoveries worthy note I shall find a proper place and repeat it. The Fly in the plate at (d) was drawn and Coloured from a beautiful large Female in the Cabinett of Charles Belliard Esq; which is the finest we have in England; the underside is shewn at (e). The Caterpillar and Chrysalis were drawn from those of M. Rosels who gives the following account of this Fly.

'This Caterpillar which is the largest of this Class, says he, is called the Sociable Caterpillar because never found alone, but always to be met with in company with others of its kind, they feed on all kind of Willow-Trees and exist all the Summer, the reason why they are scarcer than others of this Class, is because all kind of weather does not agree with them.

'The Females lay their Eggs close together on the Branches of the Willow, tho he never could find any in their maturity but found only the empty husks instead of the full Eggs and the young brood near those, in their first Skin they appear of a Dark Brown and lie all close together on a web, each of them in creeping spins a Web whereon it sustains itself, the which serves them likewise for a Ladder or Bridge to transport them from one Leaf to another.

'When they are full fed as at (a) the body is garnished with short grey Hair or Down, the Prickles on the back are about a quarter of an Inch in length and appear branched. The Head is formed like a Heart and is covered with very small blunt Points, which produce or exhibit a languid Gloss. On the back appear eight or nine square Orange-Coloured Spots. When they are near the time of their Transformation they retire to a place of Shelter from Storms and Sunshine, there fixing their hind Legs by a Web with their Heads downwards bent towards their Belly as at (b) and in a Day's Time or something longer the Skin slips off and the Chrysalis appears as represented at (c) they hang in that state about fourteen Days and then produce a Butterfly.

'When the time approaches for the Females to lay their Eggs, they seek a secure retreat among the Willows, and lay their Eggs on the Branches, if it happens to be in the warm part of the Summer, they will be hatched in two or three Weeks Time; but should it happen late in Autumn they remain in the Egg the whole Winter, and are not hatched till the following Spring: the Male and Female die soon afer Copulation even in the same Year they were produced, tho' some are thought to live, and conceal themselves during the Winter.'

The SILK-WORM
The MULBERRY-TREE

THE Worms, or Caterpillars, are produced by Eggs, the Shape, Colour, and Size of those at *(a)* about the beginning of May, or sometimes sooner; and if the Mulberry Leaves, which is their proper Food, are not put forth, the young Caterpillars may be fed with Lettice: They are naturally tender, and subject to several Diseases, to prevent which, those, whose business it is to keep them for the sake of their Silk, keep them very clean and dry, often removing them for Change of air, perfuming the Room wherein they are fed with Incense; they likewise suffer by Thunder, Firing of Guns or any harsh Sound; when full fed as at *(b)* they grow restless, and appear transparent, soon after they spin themselves up each in a Yellow Silken Case as at *(c)* and change into Chrysalis, which I have described at *(d)* as taken out of the Case; this happens about the middle of June. In this State they remain about eighteen or twenty Days, when the Moths appear. They copulate soon after they come from the Chrysalis, and lay their Eggs, which remain in that State during the Winter, and are hatched in insuing Spring: the Cock and Hen I have shewn in copulation: the Cock at *(e)* and the Hen at *(f)*; the distinguishing difference is that the Cock is much less, particularly in Body; and his Horns comb-like and broad.

The LARGE TYGER

THE Eggs, which produce the Caterpillar of this Moth, are laid by the Hen in about July, on or near the Plants which are proper for their Sustenance: she fixeth them fast by a glewy Matter disposing them in a very regular and beautiful Order:they are of a light Yellow-Green-Colour and appear glossey. The young Caterpillars separate as they leave the Eggs and go in search of food, which is chiefly the Great-Stinging-Nettle, tho' there are few Plants they will refuse; when the cold Weather approaches, they betake them to Places of retreat and security, and where they may be sheltered from the Rigour and Inclementcy of the Season in that motionless and sleeping State in which all Caterpillars of this Class lie during the Winter. In the Month of April they again appear, but I would not advise them to be sought for till the middle of May, when they are larger and appear in greater Plenty. They are found by beating the Nettles, which grow on Bank-sides, with a Stick, which causes the Caterpillar to role himself instantly up in the form of a Hedge-hog, and roll down the Bank to the Ground; you will now find them of many different Sizes, some appearing very small, scarcely a quarter of an Inch long, others an Inch and a half, or two Inches, the largest are generally full fed about the latter end of May, as is seen at *(h)* when each of them spins itself up in a light Grey Web as at *(m)* and changes to a Black Chrysalis, the figure of which is shewn at *(i)* the Moth comes forth in three weeks. The Hen is shewn flying at *(k)* and the Cock in a sitting position at *(l)* the Cock is generally smaller than the Hen, and his Horns broader.

Plate XIII

Articles of clothing made of real silk have always been considered the most luxurious and fashionable, and from time to time the manufacture of silk from the cocoon of the Silk-worm Moth, Bombyx mori *has been established in England. It was in China several thousands of years ago that silk was first made in this way; the process was kept secret, but the story relates how, in the year 552, two monks from India smuggled some eggs of the moth out of China, hiding them in a hollow cane. The eggs were taken to Constantinople where the industry was started, from whence it spread thoughout Europe and the rest of Asia, sometimes employing large numbers of people. In England it apparently flourished in Harris's time. Some fifty years earlier Albin had also described the procedure for rearing these caterpillars, commenting, as Harris did, on their sensitivity to loud noise such as gunfire or thunder, and the use of incense to purify the air. So many years of confinement resulted in these moths losing the power of flight and they can no longer be found in the wild state.*

'All diamonded with panes of
quaint device,
Innumerable of stains, and
splendid dyes,
As are the Tiger Moth's deep
damask wings'.

Keats

The Garden Tiger, Arctia caja, *which Harris called the Large Tyger is such a striking and gaudy moth that few people will not recall having met with it at sometime in their lives. The caterpillar too demands attention because of its long tufts of hair which, as can be seen in the plate, cover its body and which give it the name, so familiar to schoolchildren, of Woolly Bear. It can often be spotted scampering across the path in the early summertime, looking for a place to pupate.*

To the Right Honourable,
This Plate is most humbly Dedicated

Countess of Ailesford
by her Ladyship's most obliged & obedient Serv.t
Moses Harris

To the Rt. Honble. Lord Wentworth ~ this Plate is most humbly Dedicated by his Lordships most Obliged ~ and Obedient Servant Moses Harris.

PENSES · A · BIEN

Plate XIV

The NUT-BEETLE
The WHITE FILBERT

Corylus Sativa, fructu, Oblongo rubente, Pelicula Alba tecto. C.B.

THE Caterpillar of this Beetle is produced by a very small Brown Egg, carefully laid, and fastened by the Parent to the Hazle Nut, at the Time when the Nut is very young, and the Coat, or Shell, very soft and tender. When the young Caterpillar, within the Egg, is complete by Nature, for which there is no Rule or Determination by Time, it eats its Passage through the Shell into the Nut, without spoiling the outward form of the Egg, which, for the present, still covers the Hole or Puncture. The chief of his Food is now the Coat, or that Part which hardens and in Time becomes the Shell of the Nut. This he feeds on, together with the inward Pulp, till such time as the one becomes too hard, and the other too dry for his Sustenance: He then begins on the Nut-Kernel, which had he done when the Kernel was very small, he might possibly have destroyed that which is his future Dependance, and which is all that is allotted him by Nature for Food while in the Caterpillar State. While feeding he still has Regard to the Hole, keeping it open by gnawing away the Sides, cunningly keeping it round and smooth; which serves to give him Air, and throw out Part of his Dung or Excrement when he wanteth Room, as well as for his Passage out when Full Fed and ready for his Change, which happens in September or later; and the Nut being ripe and dry, having fallen to the Ground, he then works himself through the Hole of the Shell, which he is some Time about, as the Hole is much less than the Circumference of the Caterpillar; when out, he bury's himself in the Earth and changes to the Nympha, in which State he remains during the Cold-Weather; and about the beginning of May assumes the Form of the Parent Beetle. I have given the Draught or Representation of the Hazle-Nut at *(m)* with the Hole where it is generally made by the Caterpillar, of this Insect. At *(h)* the Nut is broke open, and the Caterpillar discovered within, which is shewn at *(i)* in full Proportion. The Nympha is seen at *(k)* and the Beetle, which it produceth, is exactly described at *(l)*.

The SCARCE VAPOURER

THE Caterpillar of this Moth is taken by beating the Oak and Hazle on which it feeds; it is generally Full Fed towards the latter end of May, and changes to the Chrysalis, within a Spinning, against the Body or Branches of the Tree, and the Moth is produced in about Eighteen Days after. The Cock is easily known from the Hen, even in the Caterpillar State, by his being so much less than the Hen, as may be seen by comparing the Caterpillar described at *(b)* and *(e)*, but the difference is much more perceptible in the Moths, for the Hen, tho' much larger than the Cock, has no Wings, nor can it at first be perceptible that she hath any Legs, but appears altogether a shapeless Lump; she appears very dull and unwilling to move, nor ever moves far from the Web, wherein she lay when in Chrysalis, but sits waiting on the Branches, or on her Web for the Cock, which has the Faculty of smelling her at a very great Distance, some scruple not to say half a mile; however, it is most certain, they do smell them a great Way off. There are many Species of the Moth Kind beside this that find their Hens by the Scent: and it is the Practice of Aurelians to take a Hen Moth, in a Box covered with Gause, to a Place near which 'tis likely there is a Brood, on Purpose to catch the Cocks, which, in a small Space of Time, come flying about the Box, and are easily taken in your Net; this is called sembling, of which I shall have occasion to speak more of, in its proper place. She lays her Eggs on the Branches, covering them over with fine Hair or Down, as appears at *(a)*. I have given the Figure of the Hen Moth at *(f)* where she is represented in Copulation with the Cock, whose upperside is seen at *(d)* and the under at *(g)*. The Chrysalis of the Male Moth is seen at *(e)*. There are two Broods of the Scarce Vapourer in a Year. The first appears from the Eggs in Spring, becomes Full Fed the latter end of May, and produces their Moths about the middle of June; and the Caterpillars, proceeding from the Eggs of this Brood, are generally Full Fed about the latter End of August, or the beginning of September; and after laying in the Chrysalides about eighteen Days, the Moths appear. The Eggs of which continue during the Winter, and produce the aforementioned Spring Brood. The last Hen Caterpillar of the Scarce Vapourer that I got, was beaten off the Sallow in Comb-Wood, the twenty-seventh of May 1760; on the thirty-first of the same Month it spun a pretty extensive Web, wherein it changed to the Chrysalis, and the Moth was produced the seventh of June, lying the usual Time for the Hens of both the Scarce and Common Vapourer to lay in Chrysalis. Now by reason of their scarcity, I though it worth my while to semble it; to which End I took it out with me in a Box, and when I came to the appointed Place, which was the little Cut in Oak-of-Honour-Wood, I took it out of the Box, and pinned it up against a small Tree, hanging by her own Web, and in less than five Minutes, two Cocks came flying down to her, one of which I took with my Fingers, and the other I suffered to join with the Hen; when they had been in Copulation some small Time, I took them both as they was, and put them in the Box with the Web, to which they both seemed to cling very fast; the next Morning I found them separated, and the Hen had laid about three hundred Eggs.

The MOTTLED UMBER

THE Eggs are laid in regular Order on the Branches of the Oak, Hazle, and Black-thorn, on which the Caterpillars feed, tho' they are found most plentiful on the Oak, and are taken by beating it, the latter End of May, when they are about full fed; they bury themselves in the Earth, where they change to their Chrysalides, after they have each of them spun themselves up in a thin Web, the Moths appear in October. The Caterpillar is shewn at *(n)* full fed, and at *(o)* it appears in an odd Kind of Posture, in which they are frequently seen, and so stupid and insensible are they at that Time, that I have found them in my Net when beaten from the Tree in that Position, and tho' I have handled them pretty roughly, they have still retained that Posture without Motion. The Chrysalis is seen at *(p)*; the Hen is shewn flying at *(q)*, displaying her upper Side; the upper Side of the Cock is seen at *(r)*. The Caterpillars and Moths vary from each other in both Colour, Size, and Markings, altho' of the same Brood, and oftentimes so much, that none but those who are well experienced would know them to be the Mottled Umber.

The YELLOW EGG PLUMB
The YELLOW or WINTER TUSSOCK

THE Caterpillar feeds on the Oak, and several Sorts of Fruit-Trees, on the Branches of which the Hen Moth lays her Eggs in Clusters, about the latter End of May; the Caterpillars proceeding from these Eggs become Full Fed about the End of September, and are taken by beating; they change to their Chrysalides, each within a light Web as at *(g)*, and the Moths appear the Middle of May.

The Caterpillar, which produces the Cock Moth is shewn at *(a)*, he is somewhat less, and is more upon the Orange-Colour than that which produces the Hen; his Chrysalis is represented at *(b)*, and the Cock Moth which it produces is seen at *(c)*. The Caterpillar at *(d)*, is that which changes to such a Chrysalis as is shewn at *(e)*, and produces the Hen Moth which is exactly described at *(f)*.

The GREY DAGGER

THE Caterpillar of the Dagger is produced by a small light green Egg, laid by the Female the Beginning of June. The Food of the Caterpillar is Willow, and most Sorts of Fruit-Trees and Shrubs, and are generally found in Gardens; they become Full Fed the latter End of August, or Beginning of September, and are then of the Size and Colour of that represented at *(h)*; they then spin a pretty strong Web on the Surface of the Ground, or in the Holes and Cracks of the Barks of Trees, Walls, &c. where they change to Chrysalides of a reddish brown Colour, as shewn at *(i)*, and the Moths are produced the May following. There is no very distinguishing Difference between the Cock and Hen Moths; the Cock indeed is somewhat stronger marked, and not quite so large as the Hen, which I have shewn flying at *(k)*, and the Cock settled on a Leaf at *(l)*.

The ARGENT and SABLE

THIS Moth which is esteemed a Curiosity, is very scarce, nor has any Body been so lucky as to discover either the Caterpillar or Chrysalis, which is the more difficult as the Moth is seldom seen, all that I can inform the young Aurelian is, that they are always taken in Woods the Beginning of June. That in the Plate at *(m)* shewing its upper Side, was drawn from a very fine one, which was taken in Cain-Wood the sixth of June. Another I took in Company with Mr. Drury, at Brentwood in Essex, the fourth of June 1760, which is the only one I ever saw flying.

They were once taken in Plenty by Richard Guy, Esq; some few Years ago in a Wood that parts Winchmore-Hill and Southgate in Middlesex, the Beginning of the same Month, but I do not hear that they were ever taken in any Plenty before nor since. The under Side of this Moth is marked the same as the upper, for which Reason I though it needless to give another Figure of it.

Plate XV

The Yellow or Winter Tussock, now known to entomologists by the common name of Pale Tussock, Calliteara pudibunda, is fairly common over England and Wales. The caterpillar is exceptionally beautiful, sporting, as may be seen by the plate, deep velvety black between the tufts on the back. Not many caterpillars were given popular English names by aurelians in the past but this one was known as the 'hop-dog' when found in the hop-gardens, for it feeds on hop as well as on many different trees. Because of the differences in the colour of the caterpillars, Harris was again mistaken in his conclusion that they were male and female forms; in fact they may appear in either a yellowish or greenish colour but this has no sexual significance.

The Grey Dagger, Acronicta psi, which is fairly common over most of the British Isles, earned its English name from the dagger-like mark on the forewings.

The pretty little black and white moth, the Argent and Sable, Rheumaptera hastata, was undoubtedly as local in distribution in the eighteenth century as it is now, although it does occur in pockets over most of the British Isles. It is a woodland species which enjoys the bright sun and warmth of early summer. The wood where Harris took the moth, in company with his friend Dru Drury, was at Brentwood in Essex. It was sure to have been a birch wood for this is the only food plant of the caterpillar, although a slightly different northern race feeds on bog myrtle. The dark olive-green caterpillar has a black line along the back, and the sides are marked with ochreous. It has the habit of spinning together the birch leaves at the end of a twig, forming a little tent from the inside of which it feeds during July and August. The chrysalis, which was also unknown to Harris, passes the winter in that state.

PL.XV.

To the Right Honourable *Countefs of Stamford*
This plate is humbly Dedicated, by her Ladyfhips, most Obedient & Faithful Serv.ᵗ
Moses Harris.

PL. XVI.

To the Right Honourable *Lord Carysfort*
This Plate is most humbly Dedicated by his Lordships most Obed.t & Obliged Serv.t
Moses Harris.

MANUS HÆC INIMICA TYRANNIS

Plate XVI

The GLANVIL FRITILLARIA
The LONG-LEAFED PLANTAIN

The Glanville Fritillary, Melitaea cinxia *is a scarce little butterfly found only in the Isle of Wight but clearly this was not always so. A steady decline in its numbers seems to have set in over the last century, for at one time it certainly occurred in several southern parts of Britain, even in places as close to London as Tottenham and Dulwich, where it was taken by James Dutfield and others. Lewin, in* The Insects of Great Britain *(1795), referred to it as the Plantain Butterfly but mentioned that it was not common. First recorded in Britain by James Petiver in* Gazophylacium Naturae et Artis *(1702), the butterfly acquired its common name from the 'ingenious Lady Glanvil' whose story as related by Harris, has intrigued natural historians for many years. The diligent researches of several modern authorities have at last solved the identity of 'Lady' Glanvil' — but the authenticity of Harris's tale has not been proved. The will of Eleanor Glanville (c.1654-1709) — not 'Lady' Glanville; the capital 'L' was just a misleading printing vagary — was certainly challenged but Moses Harris's memory was definitely at fault in saying that Ray was at the trial; he was, after all, recounting an incident which took place a half-century earlier. No reference has been traced yet to 'Miss Glanvil's Flaming Iris'.*

Parasemia plantaginis, 'called by some the Wood Tyger', is the name by which Harris's Small Tyger is known today. What a scene may be evoked by the thought of Harris accoutred as in the engraving, with his buckled shoes, breeches and tricorne hat, the flaps of his pockets and pincushion swinging in the wind, as he runs as fast as he can to capture with his clap-net the fast-flying Wood Tiger in the afternoon sunshine.

THE Female Fly lays her Eggs on the Leaves of the Plantain, to which they are fastened by a gummy Consistence. These Eggs are hatched in about fourteen Days, and the Caterpillars keep sociably together the remaining Part of the Summer; during the Winter they cover themselves over with a fine thin Web, which however the Rain is not able to destroy. At the Approach of the Spring, when the Sun gives a kindly Warmth, they come out of their Web, and seek their Food; each Night they retire to their Web, and lie altogether on the Top of it, for they seldom or ever get under it again after their first coming out. They are a very tender Creature, and never are seen to eat, nor scarcely move, but when the Sun shines upon them, and then they commonly feed all the while; but if the Sun should chance to hide his Head by the Interposition of a Cloud, they presently cease. As the warm Weather approaches, and they increase in Size, they separate themselves, wandering in search of their Food, which is the long-leaved Plantain, and such as grow in or near Woods; they are so remarkably timorous, that should you stir the Plant they are on, tho' never so little, or even tread within two Feet of it, they instantly roll themselves up in the Form of a Hedge-hog, as appears at *(b)*, and fall to the Ground, where they lie in that curled-up Form, 'till they think the Danger past. They become full fed as shewn at *(a)* toward the latter End of April, when they hang themselves up by the Tail, and change into the Chrysalis, as represented at *(c)*, the Fly appears in fourteen Days after. The Male is represented flying at *(e)*, and the Female at *(d)*, shewing the upper Side of their Wings; the under Side of the Female is seen at *(f)*, which differs very little from the under Side of the Male.

This Fly took its Name from the ingenious Lady Glanvil, whose Memory had like to have suffered for her Curiosity. Some Relations that was disappointed by her Will, attempted to let it aside by Acts of Lunacy, for they suggested that none but those who were deprived of their Senses, would go in Pursuit of Butterflies. Her Relations and Legatees subpoenaed Dr. Sloan and Mr. Ray to support her Character. The last Gentleman went to Exeter, and on the Tryal satisfied the Judge and Jury of the Lady's laudable Inquiry into the wonderful Works of the Creation, and established her Will. She not only made the Study of Insects Part of the Amusement, but was as curious in her Garden, and raised an Iris from the Seed, which is known to this Day, by Miss Glanvil's Flaming Iris.

The SMALL TYGER

THE Caterpillars of the Small Tyger, are produced by small round shining Eggs of a light-green Colour, which the Parent Moth fixes to the Food, about the latter End of June, in beautiful and regular Order, as appears at *(l)*. The Caterpillars separate as soon as they appear from the Egg, ranging in search of Food, which is chiefly Chickweed and Nettles, tho' like others of this Class, almost any Herb is welcome to their devouring Chops. They remain in the Caterpillar State during the Winter, and appear again very early in Spring.

About the latter End of April or Beginning of May, they may be taken full Fed as described at *(g)*, and are to be found in such Places as I have already directed for the search of the Cream-spotted Tyger, which are to be taken full fed at the same Time. When the Caterpillar is fit for its Change, it spins itself up in a Grey Web, as seen at *(m)*, wherein it changes to a black Chrysalis, covered with a Pearl-coloured Bloom, which I have shewn at *(h)*, as taken out of its Web. The Moth appears in twenty-one Days. I have shewn the Female flying at *(l)*, and the Male at *(k)*. They may easily be distinguished one from the other; the Body of the Hen is of a bright-red, while that of the Cock is of the Colour of Yellow Oker. They likewise differ in the Horns and Markings of their Wings, which is more easily described in the plate.

They always fly in Woods, for which Reason it has by some been called the Wood Tyger; their Time of Flight is generally after Noon about Three or Four o'Clock; they fly very swift, and commonly settle on the Ground among the Grass; if you intend to take them, you are carefully to observe the Place where they settle; then run as fast as you can, and cover them with your Net, for they are very timorous, and soon apprized of Danger, and should you approach slowly, they would have Time to disengage themselves from the Grass, in which their sudden Alarm caused by your swift Motion, but the more confuses and intangles them.

The LACKEY

The Common Province Rose. *Rosa Provincialis rubra*. Park.

THE Food of this Caterpillar is Blackthorn, Whitethorn, Brier, and almost all Sorts of Fruit Trees. The Female lays her Eggs toward the latter End of July, placing them round a Twig in beautiful Spiral Order, as seen at *(a)*. In this State they remain during the Winter, and produce their Caterpillars in the Spring following; when they proceed from the Egg, they spin themselves up together in a Web, in which they generally hide in the Night, and come forth in the Day to feed. They seem very fond of lying close together on the Top of the Web, the Sun shining full upon them, and I have observed them at this Time to have an odd Sort of Motion with their Heads, which they move at the same Instant of Time together, as if one Soul had animated their little Bodies. When they arrive in their last Skin, they are about two Inches long, and the Stripes of different Colours down their Backs, very much resemble Livery Lace, from which they are called the Lackey; when full fed as at *(a)* they stray from each other, to find a Place proper for their Change, which happens about the Middle of June, each Caterpillar then spins himself up in a double Case, as seen at *(c)*. The out-side Case or Web is not so strong, nor the Silk so close spun together as the inner one, which appears thro' the other in the Form of an Egg; when he has compleated his Web, he works a fine Powder thro' the inner Case, which has very much the Appearance of Flower of Brimstone; soon after this he changes to the Chrysalis which is described at *(d)*, as taken out of the Web; the Moth appears in about thirty Days after.

I have shewn the Female flying at *(e)*, displaying the upper Side of her Wings, she is paler than the Cock, which is of a deep Brown, or Fox Colour, as described in the Plate at *(f)*.

The CREAM DOTT-STRIPE

THE Eggs of this Moth are of a light green Colour and shining, which are fixed by the Hen on the Food, and mostly on the under Side of the Leaf in pretty regular Order, as appears at *(g)*. The Caterpillars when first hatched are almost white and hairy, but at the shifting of every Skin, they grow darker; there is very few Herbs they will not eat. They are particularly fond of Jerusalem Artichokes, and all Sorts of Plants of the Sallad Kind. They become full fed as described at *(h)*, about the Middle of August. They then spin themselves up in a light Web on the Ground, or in some Hole or Crannie of the Wall, where they change to a black Chrysalis, as seen at *(i)*, and the Moths appear in the Spring following, about the Middle of May; they are very commonly seen about this Time, sitting against Walls, Pales, Houses, &c.

The Hen is shewn sitting on a Leaf at *(k)*, and the Cock in a flying Position at *(l)*, he differs from the Hen in the Broadness of his Horns, is of a deeper Colour, and has a greater Number of black Spots on his Wings.

The ROSE MAY-CHAFFER.

THE Worm or Caterpillar which produceth this Beetle, feeds under-ground, most commonly at the Roots of Trees, and never appears upon the Surface, unless disturbed by digging, or some other accident, which is the Reason why they are so little known. They are very prejudicial to Gardeners, by destroying the Roots of Plants and Flowering Shrubs, on which they delight to feed. The Hen Chaffer lays her Eggs about the middle of June, for which Purpose she buries herself in the Ground where it is soft and light enough to admit her; where making a proper Receptacle, or hollow Chamber, she disburthens herself of her Eggs; she then returns to the Surface, and flies away, to enjoy the remainder of her Days, which generally find a Period in less than two Months after. The young Caterpillars proceed from the Eggs in about fourteen Days, and go in search of Food, which is not far from them; for the provident Mother always takes care to place her Eggs where the young Brood may have a supply as soon as they proceed from the Shells; when they have gathered Strength, they begin to Burrow different ways in search for Roots; in Winter they eat but very little, if any thing, and retire pretty deep in the Ground, to avoid being bound up by the Frost. They shift a Skin every Year, and become full fed about the fourth Year, at which time they appear of a Cream Colour, their Head and Feet brown, as appears at *(a)*. About March the Caterpillar makes a Case of Earth very near the Surface, which Case is about the Size of a midling Walnut; within this he changes to the Chrysalis or Nympha, as is shewn at *(b)*, in which he continues till the beginning of May, and then produces a Chaffer in the likeness of its Parent. When they first come from the Chrysalis, they are of a light greenish Colour, and tender, and are many Days before they come to their proper Hardness and Colour; their Food is various, sometimes they are found eating the Leaves of Plants, and often feeding within the Heart of the Rose; they are very fond of any thing that is moist; I took one once sucking the Juice that flow'd from the Bark of the Willow Tree. The Hen is of a beautiful deep green Colour, as appears at *(o)* and *(p)*, which last I have shewn flying, to display the transparent wings. The Cock appears of the Colour of burnished Copper, when it is a little tarnish'd, tho' in some positions it hath a greenish Cast; I have described him folding up his Wings at *(q)*.

Plate XVII

The fanciful name used by Harris for the moth Malacosoma neustria *would not have been so clear to us without his explanation. However, the 'Lackey' of the early aurelians, which reminds us of the costume of a different age, is fortunately with us still. Harris's observations on the earlier stages of this moth, particularly the sunbathing habit of the caterpillars and the double cocoon of the chrysalis, are most perceptive, but he may not have known that the 'Flower of Brimstone' dust was first ejected by the caterpillar as a fluid, which then forms the powder upon drying.*

The Buff Ermine, the scientific name of which is Spilosoma luteum, *is certainly easier to remember and just as attractive as Harris's name. It was so christened by later aurelians because of the supposed resemblance of the black-flecked wings to heraldic ermine.*

The Green Rose-chafer is a beautiful beetle, described accurately and sympathetically by Harris. In the last century it was regarded as a most undesirable pest, for not only does the larva attack the roots of many plants, but the beetle can be exceedingly injurious, destroying the flowers of strawberries, turnips and other crops, as well as garden plants. John Curtis in his book Farm Insects *(1860) describes how they may easily be caught with a ring or bag-net when the sun shines, giving detailed instructions for the manufacture of this equipment. But for all that, it is a most attractive insect! Its scientific name is* Cetonia aurata. *Harris called it* Scarabaeus horticola *but* horticola *applies correctly to the Garden Chafer, a beetle with reddish-brown wing cases.*

Mr Harris ad Vivum.

To y.^e R.^t Honble Lady Carysfort *This Plate is*

Most Humbly Dedicated by her Ladyship's most Obliged & Obed.^t Humble Servant.

MANUS HÆC INIMICA TYRANNIS. Moses Harris.

Mo. Harris ad Vivum Se.

To the Honourable   *Tho.ˢ Townshend.*

This Plate *is most humbly Dedicated by* His   *Obliged & Obedient Servant*

Moses Harris.

Plate XVIII

The PEPPER'D.

The smooth leaved Elm. *Ulmus folio Glabro*. GER. Fm.

ELM, Lime, and Oak, is the Food of the Catterpillar which produces this Moth. The Hen lays her Eggs on the Food, about the middle of June, which stick fast where they are laid. When the young Caterpillars appear from the Eggs, they stray from each other in search of Food. They become full fed the beginning of September, when they appear of several different Colours, some being almost black, as is shewn at *(a)*, others appear of a very light Olave, like that at *(b)* sometimes they have little risings or protuberances in several Parts; when full fed they come down the Tree and burrow into the Ground about four or five Inches deep, where they change to black shining Chrysalides, having a sharp Point at the Tail, as shewn at *(c)*, which is the largest Chrysalis, and produces the Hen Moth; and that at *(d)* the Cock. They remain in Chrysalis during the Winter, and the Moths appear the latter end of May. The Hen is shewn flying at *(e)*, and the Cock in a setting Position at *(f)*. The Catterpillar is taken by beating the Trees, and the Chrysalis by digging at the Roots of Trees in the Month of April, with a Trowel. The Moths are sometimes found sitting against the Bark.

The RED UNDERWING.

THE Caterpillar which produces this Moth is seldom seen, which is owing to the Method he takes to secure himself, for he never ventures to the Ends of the Branches, but carefully keeps on the larger Arms near the Body or Bole of the Tree, from which place he can never be shaken by our method of beating. In bad and windy weather he comes down to the body of the Tree where he hides himself in the Cracks or Furrows of the rough Bark; so that by the flatness of his Make, and his Colour being so much like that of the Bark, he eludes the search of our Eyes. They generally become full fed the beginning of July, when they appear as at *(g)*, of a light olave Colour, having eight middle Legs, two holders behind, and fix crooked Claws near the Head. It now wanders about the Bark to find a Hole or Crack wherein he may be concealed during the Time of his being in Chrysalis State; but the Place they most approve of is under the Bark, or between that and the Tree, where they change within a tender Web. The Moth appears in twenty one Days after. I have shewn the Chrysalis as taken out of the Web at *(i)*, which is covered with a fine Bloom; the Moth is shewn at *(k)* flying, to display the upper Side of his Wings, at *(l)* to discover his underside, and at *(m)* in the Position in which he sits against the Body of the Willow Tree, where they are often found. They fly in an Evening, and sometimes in the Day-time, if the Weather prove hot.

Plate XIX

The LIME SPECK.
The AFRICAN MARYGOLD.

THE Hen LIME SPECK lays her Eggs about the Beginning of July, on the underside the green Leaf of the African Marygold; they are exceeding small, of a light green Colour, and placed in order like those of the TIGER Tribe. In this state they remain about a Fortnight, and then produce a small green luper Caterpillar, which feeds on the green Leaves, yet appears to be particularly fond of the Flower; under which they conceal themselves while they feed on its leaves. They become full-fed towards the latter End of August, and are about an Inch in length, they are of a fine green Colour, and variously spotted, as is shewn at (a) and (b); others appear only of a dark green, having no Spots or Markings as seen at (c). When ready for their Change they retire down the Plant to the Earth, in the Holes or Cranies of which they spin a Case, or rather Web, wherein about two Days after they change to their Chrysalides: The Colour of the Tail of which is brown, and that part which contains the Wings of a fine green. They remain during the Winter in the Aurelia or Chrysalis state, and produce their Moths the latter End of June following. The Hen is shewn flying at (d) displaying her upperside, and the Cock as settled on the Box at (f). They fly in an Evening in Gardens.

The CRIMSON UNDERWING.

OF all the Caterpillars which are attained by beating, perhaps none is acquired with so much Trouble and Fatigue as the Caterpillar of the CRIMSON UNDERWING, and their great Scarcity added to their Beauty makes them be held in great Esteem among the Aurelians. This beautiful Insect feeds on vast and high timber Oaks, whose Arms and Branches extend a long way from their Bodies, and never on the lower and younger Trees. For this Service your Sheet should at least be seven Yards long, and five broad, your Prowler likewise proportionable, which must be near sixteen or eighteen Feet long to reach even the lower Branches of such Trees on which these Caterpillars feed. Your Box, or rather Cage, must have convenient Places in the side cover'd with Gause or Crape to admit of Air, that the Caterpillars may not be over heated; in the bottom of the Cage a Bottle should be fixed full of Water with a Sprig of Oak in it, the Stalk of which should fit the Mouth of the Bottle very nicely, that the Caterpillars may not get in and drown themselves, which they will do if they can get at the Water; a spare Box or two should always be taken in your Pocket, one for Flies or Moths, and others for Caterpillars, which you may beat or chance to find on other Growths: The Caterpillar Boxes for the Pocket should have an oblong Hole in the Lid, and such a thin Brass sliding Cover as is seen in the Plate. The Time of beating for them is the latter End of May, or from the twenty-fifth of that Month to the sixth of June, when they are full fed, and appear like that at (g) which is a just Representation of the Caterpillar. It is about three Inches long, and thick in Proportion, having several Lumps, or Protuberances on the Back, out of which spring several small Hairs; they are variously mottled or chequered with a Variety of Shades, the Sides, near the Belly which is flat, are furred, or rather fringed, with short downy Hair. It is of so odd a Colour, that, if you are not very careful when you are beating for them, you will be deceived when they are in the Sheet, and throw them away together with small Bits of broken dead Twigs; for those trees are generally covered with a rugged greyish Moss, not unlike the Caterpillar, even when viewed pretty nearly.

When about to change they search for some convenient Hollow, or Cavity, in the Bark, or hide themselves among the Moss, with which the Arms and Branches of such old Trees greatly abound; here they spin a brown Web of a middling Strength, wherein after a few Days they change to the Aurelia or Chrysalis State, which I have described lying on the Box at (h); they are of a deep red brown Colour, having that part which contains the Wings covered with a Bloom, which moves or comes off when touched; they remain in the Chrysalis State about three Weeks, so that the Moths appear the beginning of July.

There is a very great Resemblance between this Moth and the RED UNDERWING; their Chrysalides are also formed alike, and both covered with Bloom; the Caterpillars are both flat on the Belly with a round black Spot on each of the six middle Joints, and the Caterpillar of the CRIMSON UNDERWING hath much of that grey, or light Olave Colour, which covers the Caterpillar of the RED UNDERWING. I have shewn the Moth flying in three different Appearances, viz, at (i) (k) and (l), that a better Idea may be had of the upper and underside of this beautiful Insect. With respect to the Eggs they are no doubt laid on the Food; but when hatched, or in what State this Insect lies during the Winter, is what I cannot assert as a known Fact; but as it is observed, that all Moths, hitherto known, copulate and lay their Eggs soon after their Appearance from the Chrysalis, and this Moth being produced in the middle of the Summer, it is not doubted, I presume, but the Eggs are hatched the same Summer in which they are laid, and consequently live through the Winter in the Caterpillar State.

In this plate Harris has chosen to display a very small moth and a very large moth. The first of which he gives an account is now called the Lime-speck Pug, Eupithecia centaureata. It is one of a genus of small moths known as the 'Pugs' (the ancient name for monkeys) and is a translation of the scientific name. The popular name, Lime-speck, is supposed to derive from the appearance of the moth when resting. Widespread and usually common in Britain, the caterpillar feeds on the flowers of many plants, not only the African marigold which was the food plant upon which Harris's 'hen' laid its eggs.

The Dark Crimson Underwing, Catocala sponsa, the Crimson Underwing of Harris, at first glance looks very similar to the moth shown on the previous plate. However, it will be noted that the forewings are darker and the hindwings a richer crimson; and the larval food plant is quite different as this one feeds on oak exclusively, which the Red Underwing avoids. Furthermore, this is a very scarce insect, inhabiting, so far as known, only the oak woods of the New Forest in Hampshire. But Moses Harris and his fellow aurelians knew it, although he said it was scarce. Regrettably he gave no locality where it could be obtained at that time, although it was almost certainly in the London area. His assumption that it hibernates as a caterpillar was incorrect, as the winter is spent as an egg. Harris's account of collecting and rearing the rather odd-looking caterpillar is particularly fascinating to the entomologist who is interested in the equipment used by aurelians in the eighteenth century. The 'prowler' is assumed to be the long pole with which the branches were agitated to dislodge the caterpillar and make it fall on to the huge sheet spread out below. This apparatus is incorporated in the little vignette engraving on the title page. That he was aware of the importance of plenty of air for his captures, and the necessity of making certain the mouth of the water bottle was 'very nicely fitted' with the oak sprigs, demonstrates the care taken by this artist-aurelian. The roomy 'caterpillar box for the pocket', with the sliding brass cover over the oblong hole, appears to be a useful innovation.

To His Grace *Cha.ˢ Lenox* *Duke of Richmond,*
This Plate is humbly Dedicated by his Grace's most Obliged & Obed.ᵗ Hum.ᵇˡᵉ Serᵗ
Moses Harris.

Pl. XX

Plate XX

The alternative name for the Lime Hawk-moth, the Olive Shades, unfortunately went out of favour many years ago, although it is much more attractive than the prosaic one now in use. Mimas tiliae is a moth which is subject to colour and pattern variation in the wings, and is common over southern England, particularly in the London area. Nevertheless Harris was left with a few doubts about the economy of the insect. He rightly queried the food plants of the caterpillar, for it will feed also on alder and birch, and his opinion that it spun a little web in which to pupate, although he was unable to confirm it, was correct. However, his observation that this took place four or five inches deep in the ground was unusual, as an inch or two beneath the surface, or even on the tree-trunk, is more commonplace. The plate is particularly instructive to one interested in the tools and method of setting butterflies and moths, which he explains at length in his Introduction. The procedure is remarkably similar to that in use today, except that transparent strips are now employed in lieu of the card braces, and the English fashion is now to display more of the hindwings; the refinement of a groove in the setting-board to accommodate the body of the insect did not develop until a later date.

The Vapourer moths, the Scarce, shown on Plate XIV, and the Common, Orgyia antiqua which Harris also called the Limetree Tussock, were given their popular English names because of the hovering flight of the males. This species may be found throughout the British Isles and is as frequent within London now as it was in Harris's day. His comment that the history of the two moths is similar is basically correct, except that this species overwinters as an egg (resembling one of the 'small Beads of a Necklace'), whereas the Scarce Vapourer hibernates as a larva.

The LIME HAWK, or OLAVE SHADES.
The LIME TREE. Tellia.

THE Caterpillar, which produces this Moth, feeds on Elm and Lime Trees, and are generally taken by beating in the month of August, at which time they are full-fed, and begin to go into the earth. I am of an opinion that Lime and Elm are not the only Trees they feed on, for I lately found one in a lane in which neither of those Trees grow, at least within several hundred Yards of the Place where I found the Caterpillar, neither can I now remember what I found it on, but Lime is supposed to be the Food it most delights in, and those Caterpillars produce the best Moths that are fed therewith. The eggs are laid by the Hen and fixed to the Food about the beginning of June, and are hatched the latter end of that Month; they are of a light shining green Colour, and a little flatted, as appears at *(a)* particularly on the underside. The young Caterpillars are remarkably long, their little tail's near the same length as their bodies, and ends in a long sharp point as fine as a hair, which appears as if dry and withered, nor do I believe they have any sense of feeling from half way to the point, for every time they shift a Skin the Tail appears shorter in proportion to the Body. They throw off five Skins during the time of the Caterpillar-state, one every seven Days, and arrive at their full Size the fortieth Day, or the middle of August, they now retire down the Tree to the Ground, wherein they bury themselves about four or five Inches deep: I never could perceive they spun any Web to form the Cell wherein they change; but I am certain the greatest part of the Caterpillars which go in the Ground do.

I was in the same doubt concerning the PRIVET, but on taking one carefully out of the Mould I discovered a fine thin Web, not unlike in Texture to the Web which a small Sort of Spider spins in the Corners of Rooms; and as the LIME HAWK is of the same Class with the PRIVET, I do not doubt but they likewise spin a Web, though perhaps so fine, that, on raising the Earth to come at the Chrysalis, it may very probably through its tenderness be destroyed. The Caterpillar lies in the Earth several Days before it puts on the form of the Chrysalis, during which time it shortens and grows thicker, particularly about the shoulder part, behind which at length it bursts the Skin, and the Chrysalis gradually appears as the Skin is driven off to the Tail by a particular Motion of the Caterpillar. The Chrysalis remains during the Winter, and the Moth appears the latter end of May, though sometimes not till the beginning of June.

Some of these Caterpillars are of a very different colour, and greatly vary from the others in respect to the Markings, some are of a light blue green, with seven oblique feint yellow Streaks on the Side as seen at *(c)*; others are of a fine beautiful yellow green, and the Streaks on their Sides yellowish, adjoining to each of which, toward the back, appears a fine Carmine or Crimson Spot, and the same, though not so strong at the bottom of each Mark toward the belly as is shewn at *(b)*. The Chrysalis described at *(e)* produces the Cock: the Hen Chrysalis only appears a little thicker.

I have described the Moth in four different Places in the Plate, the uppermost of which at *(d)* is a Hen shewing the upper side; the underside is seen at *(g)* where it is shewn hanging in the manner which all Moths and Butterflies do, when they come from the Chrysalis, to expand and dry their Wings. The Cock is shewn at *(f)* as pinned down to the Setting Board, with his Wing in their resting Position. The other Moth on the Setting Board, is a Hen with her Wings expanded; which figure is only to shew the Manner in which the Card Braces is fixed, in order to the setting a Fly or Moth, which braces together with the Setting Board and Point, is fully described in the Introduction.

The COMMON VAPOURER, or the LIMETREE TUSSOCK.

THE caterpillars are so common, that they may be found in almost any Garden in Town, for they will feed on many of the Plants that are generally used to decorate Gardens, as Sunflowers, Marigolds, &c. Those that are found wild, I mean in the Fields, feed on Oak, Elm, Blackthorn, Hazle, &c. But the Lime-tree is their favourite Food. There are two Broods a Year; the one changes into Chrysalis the latter end of May, or beginning of June, within a brown Spinning, and produce their Moths about the middle of June. The Caterpillars, proceeding from this Brood, which are full fed the latter end of August, change to their Chrysalides, and appear in the Moth State the end of September; these lay their Eggs on their Web, which continue in that State during the Winter, are hatched in the Spring, and arrive to the Moth State about the middle of June as above: In short, the whole History of this Moth entirely corresponds with the SCARCE VAPOURER: The Hens of both Species are without Wings: They have each two Broods in the Year; and the Hens appear from the Chrysalis in a Week, while the Cock remains sixteen or eighteen Days: They are taken by Sembling, otherways not easily caught.

The Caterpillar full-fed is shewn in the Plate at *(i)*. The Web, or Spinning, of the Female is described at *(h)* with the Eggs thereon, which appear perforated like the small Beads of a Necklace. The Chrysalis of the Hen is seen at *(l)* which is of a light Green or Olive Colour, and that of the Cock at *(p)* which is black and shining. They fly in the Day time in Town, hovering about Lime Trees in search after the Female.

The UNICORN.

Small Bindweed. *Convolvulus minor Vulgalis*. Park.171.

I Have taken a great deal of Pains to come at the History of this scarce and valuable Insect; for altho' many have been taken in the Moth state, yet the Manner and Places in which they were found, gave not the least Light how more were to be obtain'd. For this Reason I greatly despair of giving a satisfactory Account, but being well acquainted with the Class to which this Insect belongs, and by comparing the Accounts given me by my Friends, I am enabled to satisfy the Curious in the several Particulars.

The last Two were taken the latter End of September, 1760, in the Dusk of the Evening, by Mr. PETER COLLINSON, at his Country House, at MILL HILL in the Parish of HENDON in MIDDLESEX, one of which was caught in the Green House, and the other about half an hour afterwards, as it was beating against the Outside of the Windows, endeavouring to get at the lighted Candles. As to the Caterpillars I never heard but of two that were ever found; one by Mr. SOUTH of HAMPSHIRE, which he said was green, and appeared in other Respects so like the Privet that he was deceived: He fed it on the Leaves of the lesser Bindweed: It changed into the Chrysalis in the Earth in July; and the Moth was produced in September; this proved a Hen, and is now, together with a fine Cock, in the curious Cabinet of JAMES LEMON, Esq; from which Moths those in the Plate were drawn. The Caterpillar in the Plate at *(a)* is of the Brown sort, and copied from Marian, which is the same with Mr. ROSEL's; and, exactly agrees with the Description given me by a Gentleman of my Acquaintance, who tells me it was found in the Ground, about the Middle of Summer; but it being in the Possession of another Person, he could not inform me what became of it after. By the above Accounts we learn there are two Sorts of Caterpillars; the brown, and the green; the last of which (according to that taken by Mr. SOUTH) produces the Hen-Moth. Mr. ROSEL, tho' he has given good Draughts of both Sorts in his ingenious Work, is silent in this Particular.

I do not question but this Insect is as plentifully produced in England as any of its Class; but their Manner of Hiding and Burying themselves in the Ground, makes them Difficult to find; nor do they ever come up or appear on the Surface but to feed, which is in the Evening, sometime after the Setting of the Sun; and there is great Reason for their so doing; for as their Food runs along the Ground, (unless it meets with some Plant or Shrub in its way, and then indeed it ascends by twisting round it) and mostly on rising Places, which fronts the Mid-day Sun; the Earth, more especially at that Time of the Year, would be too hot for the Creature to exist; it would likewise be so much exposed to the Ichneumon, as this Plant affords but little Shade, that the whole Species would in Time be in danger of being Extinct; which is what Providence never admits of in any of the most trifling of her Works: This, I say, is another Reason why they should conceal themselves in the Day time; for Nature has Indued them with sufficient Instinct, by which they are sensible of an Enemy, and their Manner of attacking them which would be intire Useless, without it further Instructed them, either in a Way of Defence, or a proper Method of Securing themselves.

About the Middle of July, the Caterpillar arrives to its full Size, and Makes a convenient Receptacle in the Earth; wherein, after a few Days, it changes to the Chrysalis: In this State it continues about two Months; so that the Moth appears about the Middle, or toward the latter End of September: The Hen lays her Eggs on the Stalks of the Food, to which they are fixed like others of this Class; and in the Evening she betakes herself to her beloved Diversions. The Eggs are hatched in a few Days, and the young Caterpillars, when arrived to about their third Skin, retire into the Earth for the Winter. In the Spring they feed again, and become full fed in July as above.

The Method in seeking this Caterpillar, is, by the Dung, or Excrement, which is like that of the Privet, and must be carefully sought for under the Plant. Having found this, mark the Place, and return at Night with a Candle and Lanthorn to seek the Caterpillar; but I rather think there is a greater Likelihood of Success in seeking the Moth, which must be towards the latter End of September. The Method I propose, is to go with a Candle and Lanthorn, in the Dusk of the Evening, to some Place where the Food grows in plenty; there fix your Lanthorn, and attend it with your Nets, which you must take with you; together with Boxes, Pins, &c. If there is any Moths within several hundred yards of the Place, they will come and beat against the Lanthorn: And if you should not happen to prove Successful in the Unicorn, you would still find other Sport to Recompense you for your Disappointment. The Caterpillar is described in the Plate, full fed, at *(a)* the Chrysalis, lying on the Ground, at *(b)* the Hen-Moth, at *(c)* and the Cock, at *(d)* with his Probocis extended; which measures in the Original Moth three Inches and a Quarter.

The LITTLE GATE-KEEPER.

THE first Appearance of this Fly, is toward the latter End of April, and from that Time to the approach of Winter, is the constant Inhabitant of the Meadows. They have three Broods a Year; the First in April, the Second in June, and the Third in August; toward the latter End of these Months they are always to be found fresh and in good Condition. The Caterpillars feed on common Meadow-Grass, on which the Female Fly fastneth her Eggs; these are very Small as may be seen near the Caterpillar in the Plate, at *(e)*. They are Hatched in about six-Days, and the Caterpillars, when arrived to about their third Skin, conceal themselves in holes under pieces of Dirt &c. during the Winter Season. In the Spring they feed again, and arrive to be full fed the Beginning of April; they then fasten themselves up by the Tail, in Order for their change to the Chrysalis, which is almost White, as at *(f)* till the inclosed Fly is in Perfection, and then Appears to change a little upon the Orange or Lightish brown, which Increases still deeper till the Fly breaks its Prison and appears: This we call the First, or April Brood, from which proceeds two others, which appear in the Fly as abovementioned. There is very little Distinction between Male and Female; except the Latter being a little Larger, and the Ring or Eye at the Tip of the Wing more Strong and Conspicuous. The Upper side of the Fly is seen at *(g)* and the Under at *(h)* where it is represented sitting on the Grass with its Wings shut. They Fly very low in Meadows, Settling on the Tops of the Grass.

Plate XXI

The horn of the fabulous unicorn presented the imaginative aurelians with the title the 'Unicorn' for this great moth, now stolidly named the Convolvulus Hawk-moth, Agrius convolvuli after the food plant. The really significant fact of which Moses Harris was unaware when he wrote the delightful account of the history of this insect, so far as he knew it, was that it is a migrant from the Continent, from whence it sometimes visits the British Isles in large numbers when the conditions are right. In other years it appears hardly at all. The British climate is such that it is unable to maintain its life cycle here, although from time to time the moth has been reared from caterpillars or chrysalids found in this country; but this Hawk-moth is more often seen hovering at dusk, with its proboscis extended, over the long tubular flowers of the tobacco plant or petunia or other garden flowers. Harris was well aware how strongly this superb moth is attracted to light, and the use of a modern light-trap has enabled entomologists to detect its distribution, which is sometimes quite widespread. Let us hope the proposed evening activity of Moses Harris with the endearing picture of him fixing his 'Candle and Lanthorn' in some rustic spot, awaiting with his nets the arrival of the Unicorn was not labour in vain. But if disappointed, as a consolation he said, 'You would still find other Sport to Recompense you'. Harris clearly felt that no account of moths and butterflies should be presented to the public without the inclusion of such a striking insect. The pupa is uniquely shaped, having the 'tongue' projection which has been described as not unlike the handle of a pitcher.

This little butterfly, the Small Heath, Coenonympha pamphilus which was known to Harris as the Little Gate-keeper, is perhaps one of the most abundant species, occurring almost everywhere in the British Isles.

PLXXI

To Sir Nathaniel Curzon Bar.ᵗ This Plate is most humbly Dedicated,
by His most Obliged and Humble Servant. Moses Harris.

Mo: Harris ad Viu

To the Honble Willm Richd Chetwynd. This Plate is Most Humbly Dedicated by His most Oblig.^g & Obed.^t Servt Mo.^s Harris

PROBITAS VERUS HONOS

Plate XXII

The BURNISHED BRASS.

Archangle. *Lamium.*

THE Caterpillar of this Moth feeds on the Archangle, Nettle, and many other Plants; but Archangle is allowed to be its most favourite Food. The Hen Moth lays her Eggs, which are green and shining, on the Food, which are hatched in a few Days: The young Caterpillars feed the remaining Part of the Summer, and conceal themselves in Places of Retreat during the Winter. In Spring they feed again, and arrive to be full fed towards the Middle of July, as is seen at *(a)*. They now spin themselves up in a coarse brownish Web, in any convenient Place; but most commonly on the Food, wherein they change to a black glossy Chrysalis, as appears at *(b)*. In this State they remain about three Weeks, so that the Moths appear the beginning of August: The Moth is described at *(e)*: They are seldom taken in the Moth State.

The DARK GOTHIC.

THIS Moth appears from the Chrysalis the latter End of June, at which Time they couple, and lay their Eggs. The young Caterpillars feed but slowly: The remaining Part of the Summer, and during the Winter, conceal or bury themselves in the Ground, under, or as near as they can, to the Roots of their Food; which is Water Betony, Dock, Archangle, Nettles, &c. Very early in the Spring they come forth again to feed, and become full fed, as at *(d)* about the beginning of May: They then spin themselves up in a Web on the Surface of the Earth, and change into brown shining Chrysalides; in which State they continue one Month, and then the Moths appear. The Chrysalis is seen at *(e)*: The upper side of the Moth is shewn at *(f)* and at *(g)*. The Moth is seen flying in such a Position, as discovers the under Side. This Figure of the Moth may seem unnatural, but it is a Position in which most Moths turn themselves when going to settle on the under Side of a Leaf, Board, &c. They fly by Night, and are very seldom taken. The Caterpillars are found in the beginning of April, by searching at the bottom of the Stalks of their Food, and such of their Food as grow against the Sides of Banks; for though they seldom or ever get on the Leaves, nor crawl high on the Stalk, yet sometimes, by striking the Food, they will drop off, and roll down the Bank to the Ground.

The COMMON EVENING SWIFT.

THIS Moth is generally taken flying in an Evening in Meadows: They fly low and very swift; which is the Reason they were called by this Name. The Hen casts her Eggs from her as she flies, in great Numbers at once, which are small and black, appearing like fine Gunpowder: These drop into the Grass, and rest at length in little Holes, or Crannies of the Ground; where, after continuing a few Days, they are hatched. The young Caterpillars immediately begin to feed as near the Root as they can; and, after a little Time, being grown larger and stronger, they eat their Way to the very Roots, perhaps four or five Inches deep, where they make long and winding Passages among the Roots; and from one Knot of Roots to another: these Passages are all carefully lined with their Web, to hinder the Dirt from falling in, and interrupting their Way. In this Manner they feed and live during the Winter Season, seldom or ever coming out of the Ground, any farther than is seen at *(h) (h)*, and even then, at the moving of a Bit of Grass, or the blowing of the Wind, they will suddenly draw in their Heads, and run into their Holes a great Way backwards. They become full fed the latter End of April, or beginning of May, and appear as at *(i)*, of a Cream Colour, and their Heads red: They then come nearer, or within an Inch of the Entrance of one of their Passages, where they inclose themselves in a slight Web, and change into Chrysalides, which is of a crooked or bending Form, as at *(k)*. They continue in this State till towards the End of May, when the Moths appear. The Hen is seen flying at *(l)*, and the Cock at *(m)*: The latter appears generally more upon the Orange Colour than the Hen, and the Wings more transparent, as if the Feathers were rubbed off: But it is not very easy to discover the Cock from the Hen; so much do they differ, both in Size, Colour, and Markings.

The COMMON LADY BEETLE.

ON the 29th of May, 1759, I found on an Oak Leaf, a Cluster of pretty large Eggs, consisting of twenty-four, of a very light green Colour, and appeared as at *(n)*. I carefully brought them home, and, on examining them with my Glass, discovered them to be of an oval Form, not growing taper towards the Ends, but rather shaped like a Barrel or a Cask: I placed them on some Food, resolving to examine them every Day, expecting they would produce some strange Sort of Caterpillar, not remembring ever to have seen Eggs of that Shape before. However, I saw no Alteration in them till the 10th of June, and then there appeared on each Egg, pretty near the top, a black Mark of a triangular Figure, which was divided by another Mark, reaching from the upper Side to the lower Point or Angle, as may be seen at *(p)*; where one of the Eggs is shown as seen through a good Magnifier: Thus they remained till the 13th, and then I perceived the Eggs to be changed to a lightish brown, or Wainscot Colour; and two additional Marks appeared a little above the Triangle: These were of a fine bright Crimson Colour; and now I expected the Caterpillars to appear very soon; and accordingly next Morning found them crawling about the Box: They appeared very dark, or rather black, having no Belly-Legs, nor Holders, by which I knew them to be of the Beetle Kind. About the latter End of July, they appeared as I have represented them at *(o) (o)* full fed; when they fastened themselves up by the Tail, and changed into Chrysalides, as seen at *(q)* and *(r)*, which last is seen in a Profile position. This Chrysalis when disturbed raises itself almost erect on its Tail, with a quick and sudden motion, as shewn by the dotted line near the Chrysalis at *(r)*. They lie in the state seven Days, and then produce small Beetles, commonly called Lady-birds: the Caterpillar will feed on almost any Plant, and they lie during the Winter in their Beetle state.

The beautiful golden sheen on the forewings of the Burnished Brass, Diachrysia chrysitis was not easy for Harris to convey, and the moth must be seen in reality to be appreciated. It is double-brooded, which led Harris to assume the moths appeared in August. This is true of one generation, but the young caterpillars which overwinter are fully fed in May and so moths are produced in June and July as well. It is not rare over most of the British Isles and frequently visits garden flowers, as well as being attracted to light, so it is surprising that Harris rarely took it, but then distribution patterns may have changed considerably over two centuries.

The arches on the forewings of this delicately shaded moth, the Gothic, Naenia typica gave rise to Harris's English name which is still in use. Although it is never very common, it can be found all over the British Isles. One interesting point not mentioned by the author is the habit of the tiny caterpillars to live, before they overwinter, in companies on the undersides of the leaves of their food plants, which include a wide variety of plants, shrubs and trees.

The Common Swift, Hepialus lupulinus must be familiar to everyone who has walked through English meadows at dusk in the summertime, when this little moth can be seen careering rapidly over the grass. For a time Harris's name for the insect was superseded by the Silver Swift, used by Haworth in 1803, but now Common Swift is the familiar name. The account of the larva is an example of remarkably accurate observation.

Moses Harris made good use of his magnifying glass when studying the minute details of any tiny natural object before him, but occasionally he omitted to record something which would have added to the interest of the subsequently related account. In his remarks on the Common Lady Beetle, he did not indicate the nature of the food upon which he placed the eggs, from which the ladybirds were successfully reared. However, although the impression is given that he was unaware the little larvae fed on aphids, which are so destructive to plant life, this fact must have been known to Harris for it was common knowledge at the time. Albin, for example, described some fifty years earlier how he fed the 'Cow-lady' on small insects, and indeed this was the reason they were so well regarded.

The GOAT MOTH.

The Willow-Tree. *Salix.*

Plate XXIII

THE Hen Moth lays her Eggs on the Bark of the Tree, in a confused Heap, about the beginning of June, these are of a Dirty Brown Colour, and appear of an irregular form, as seen at *(a)*, they are Hatched in a few Days, perhaps ten or fourteen, and the young Caterpillars immediately divide and begin to feed on the Bark of the Tree, getting for the first Year (for they are three Years in the Caterpillar state,) no farther into the Tree than under the Bark; the second Year they get into the Woody and more substantial Part, eating through the very Heart of the Tree, from the Top to the Bottom of the Body or Bole, and it is very common to see the Trees on which they feed, or where there is a Brood, killed by this devouring Insect; by destroying so much of the Wood and Bark, as in Time fairly obstructs the motion of the Sap and Juices. In some places, particularly in Cornwall, they are called the Auger Worm; and indeed the holes they make appear as if bored with an Auger and so large as to admit a Man's Thumb. The third Year they arrive to be full fed, and then appear about five Inches in length, and, in my opinion, very disagreeable to the eye, appearing like a large Maggot of a pale and sickly Clay Colour: the Back is of Deep Purple-Brown; the Head is Black and Shining, behind which on the first Joint it hath a Spot of a remarkable shape, which is glossy and seems of a hard substance: from the Sides appear a few tender hairs.

Such a Caterpillar as I have now described is seen in the Plate. But many of them, especially the young ones, are much paler. When they are ready for their transformation which happens about the beginning of April, they come within four or five Inches of the Mouth of one of their Holes, where they knaw away the Wood till they have made a Cavity large enough for their purpose; here he makes a very thick and warm Case, composed of his Web and the little Bits of Wood that he has knawed off making the case appear as if covered with Sawdust, the Inside being delicately smooth and shining like White Sattin: within this case he changes to the Chrysalis as seen out of the case at *(c)*: he remains in Chrysalis about two Months, and at the expiration of that Time, being ready to come out, he drives himself very forcibly forward out of the case through a weak part which he providentially leaves to facilitate his passage; when he is delivered from the casee he still drives himself forward till he has approached very near the Mouth or Entrance of the Hole, where he immediately breaks from the Chrysalis, leaving the empty Shell half out, as is represented in the Plate at *(d)*.

BUT to return again to the Chrysalis. Perhaps not being well acquainted with this Insect, you would ask how the Chrysalis which does not appear as formed by Nature to help itself in any shape, should be capable of moving forward, not having the assistance of Legs. But if my ingenious Reader will please to refer to my Draught of the Chrysalis in the Plate, the mistery will vanish: he will there find on the back of the Tail part, every Joint or Division armed with a great number of points, or Saw-like Teeth, which first by contracting and then extending himself, these points fasten in the Case, and by such kind of Motion the Head of the Chrysalis must go forward, for the Teeth or Points cannot be relieved by any other means, and I do not doubt but the Hooks at the Head were intended to assist him at that time. When the moth is relieved from the Chrysalis and his Wings dry, He sets Himself in an odd position against the Tree, his Tail and ends of his Wings lying close and his Head or back part jutting out as if ready to fall from the Tree, thus he remains, as seen at *(e)*, till Night, when he takes wing and flies away.

The upper side of his Wings are seen at *(f)*. The under-side having a very Colourless appearance and few or no markings, I thought it needless to give another Figure of the Moth: and the Hen is much larger than the Cock and richer in Colour, the markings on the Wings of the Cock are less in Number, and its Wing more pale and Colourless. This is one of the most difficult Moths of any we have to breed from the Egg; for no Wood is proof against the hardness of their devouring chaps, therefore very troublesome to confine them, and the tiresome length of Time they take in feeding make it altogether not worth the trouble. Nor will a Glazed Pan or Jug hold them, without covered at Top with a Plate, Tile, or some hard substance; for they can with ease get up the sides by making with their Web a sort of Ladder, by which they easily ascend any steep place. See the Caterpillar in the Plate at *(b)*, which is creeping on a Ladder of this sort, and if touched or disturbed in their motion, they will turn about and bite at its opposer with great fury.

I cannot break off the history of this Moth without mentioning a Chrysalis of one of these Moths which I took out of a Willow-Tree: it was remarkably Black and soft, which caused me to think the Moth would in a short Time make his appearance, but on keeping the Shell two or three Days and not seeing any further alteration, I broke it open and to my surprize saw it so full of very minute Ichneumon's, that it appeared as if it had been crammed with Gunpowder; and to guess within a moderate Computation, I suppose there might be at least twenty thousand. And how this wretched Insect underwent its Transformation while so full of those destroying Animals, is to me very extraordinary: for it is very clear they must make their entrance while in the Caterpillar state, for the Chrysalis is too well guarded with a thick and tough case for an attack of that sort.

The fat and oily caterpillars of the Goat Moth, Cossus cossus made a luxury dish for the ancient Romans. They even, said Pliny the Elder, 'franke them up like fat-ware, with good corne-meale', according to an early seventeenth-century translation. Such a delicacy would not appeal today, for in addition to its unpleasing appearance — and Harris found it disagreeable enough — this unattractive caterpillar emits a smell which has been likened to that of a he-goat, hence the popular English name. There is no doubt too that it can cause a great deal of damage to timber, sometimes killing the trees, and an idea of the size of the holes made by these insects was given by Dr Derham in his Notes *in Albin's* Natural History of Insects *in the second edition of 1724. He had noticed in Essex, he said, 'holes of some three quarters of an inch, eaten in the timber on which the bells in Upminster steeple are hung'. Ray described the caterpillar and moth in 1710, but Harris's detailed life history has been acknowledged by entomologists to be a marvel of accuracy. It is a moth which is becoming rather scarcer.*

To the R.ᵗ Hon.ᵇˡᵉ the   *Earl of Suffolk*

This Plate is most humbly Dedicated   *by his Lordship's most Obed.ᵗ Serv.ᵗ*

NON·QUO·SED·QUOMODO

Moses Harris.

To Sir Armine Wodehouse Bar.ᵗ Dedicated by His most Obliged

This Plate is most Humbly and Obedient Servant. Moses Harris

AGINCOURT

Moˢ Harris ad Vivum

Plate XXIV

The DOTT.

The Greater Bindweed. *Convolvulus Major.*

THE Caterpillar of this Moth is commonly found feeding in the Ditches, about the middle of August, upon Nettles, Knott-Grass, the greater Bindweed, and many other Plants. There are two sorts of them, the one Green the other Brown, of which the Brown sort always produce the Cock Moth. They are full fed as at *(a)* and *(b)* about the latter end of August, when they go into the Ground and change to Brown Shining Chrysalides, like that at *(c)*. About the end of May they emerge into the Moth state; the Hens lay their Eggs soon after Copulation, which adhere to the Food, and the Caterpillars proceeding from these Eggs, become full fed about the latter end of August, as above. The Moth is seen flying at *(e)*; and at *(d)* sitting with his Wings close. They are seldom taken in the Moth state.

The HUMMING-BIRD.

THIS Moth is taken flying in the day time about the beginning of August, they fly exceeding swift, and are therefore difficult to take: but the Caterpillar and Chrysalis have never yet been found in England.

MARIANA, who has bred them from the Caterpillar, says they feed on most sorts of Plants, and that the Ladies Bedstraw is one. The Caterpillar in the Plate, with the Chrysalis, I copied from the drawing that was sent Mr. Roosel, which agrees exactly with that in Marian; by which drawings I find there are two sorts of the Caterpillars, the Green and Brown, which I have shewn in the Plate at *(f)* and *(g)*. The Figure of the Chrysalis is seen at *(h)*, the Moths I drew from one of our English Moths, the upper side of which are seen at *(i)*, and the under at *(l)*.

As with several other caterpillars in which a predominantly brown or green form may be met with, Harris's conclusions (although perhaps valid in respect of the particular ones from which his account was drawn) that these were sexual differences, cannot be substantiated. The Dot Moth, Melanchra persicariae *is fairly widespread and common over England and Wales, but the moth usually flies in July, rather later than the date noted by Harris.*

Harris's assertion that the caterpillar and chrysalis of the Humming-bird Hawk-moth, Macroglossum stellatarum *had not been found in England was made without the knowledge that this welcome little Hawk-moth is a regular visitor to Britain from the Continent, and is unable to maintain its life cycle over the English Channel. They are sun-loving moths, occasionally to be seen hovering in front of garden flowers such as petunias, honeysuckle, buddleia and valerian, and may be encountered any time from spring to autumn. Although it is regularly recorded, the numbers, as with all these immigrants, vary greatly from year to year. Occasionally their appearance in Britain causes some bewilderment as, when seen darting from flower to flower, hovering and extending the long 'tongue', into the blooms, the resemblance to a humming-bird from a more exotic land is striking. Caterpillars and chrysalids bred from the eggs laid by these guests are sometimes found in Britain but the moth rarely, if ever, survives the winter away from the Continent. It can, however, hibernate in its native climate.*

The EMPEROR.

The Blackthorn. *Prunus Sylvestris. Germ. Emac.*

THE Eggs of this beautiful Moth are generally fixed by the Hen round the Stalks of the Food, depositing them in several different places, that she may be the more certain of the security of some of them: Should these appear of a very light green they are naught, and you may be sure the Hen has not been copulated; but if of a Brownish cast as at *(a)* they are good. They are generally allowed to be a Month in the Egg state, though sometimes the Caterpillars appear sooner; but then they generally all die. The young Caterpillars are sociable, and keep together till about their third Skin, after which they separate and are difficult to find. In their first and second Skins they appear very dark, and have a Cast of Orange Colour down their backs; but when they put off their third Skin, the Green begins to appear. They shift six Skins before they arrive to be full-fed, which happens about the middle of July; when they spin themselves up in Brown Cases in the manner shewn at *(i)*, in order for their change. They remain in Chrysalis during the Winter, and the Moths appear the beginning of April: but should they not all come out then, as is frequently the case, the rest lie till that time twelve Months, and produce Moths not inferior to those produced the year before them. The Hen Moth I have shewn flying at *(g)* shewing her upper side; the under side is seen at *(h)*. The Cock is much less, and the under Wings are of an Orange Colour as is shewn at *(d)*. The Caterpillars likewise differ from each other, for that which produces the Male is stronger marked with black as may be seen at *(b)*, while that which produces the Hen is larger and fuller of green. Their Chrysalides may also be distinguished from each other, the Cock being less, and that part which contains his comb like horns very broad as shewn at *(c)*. The figure at *(f)* is an empty shell of the Hen Moth, which is to shew in what manner the Tail part of almost all Chrysalides, of the Moth kind are extended when the Moth leaves it.

The Food of the Caterpillars is chiefly blackthorn tho' they are often found feeding on the Willow or Osier in Chelsea Aytes, which is the most certain place to find them; the best time is about the middle of May when they are in the Black Skin, and as they are then sociable are more perceptible to the Eye. The Web or Case wherein they change to the Chrysalis is greatly to be admired, being so wonderfully formed for the security of the inclosed Insect; the entrance or part designed by Nature, for the coming forth of the Moth is so contrived, that it is almost impossible for any Insect to enter; and should they attain the Mouth of the Case, which they cannot do, without being very much embarrassed with the Web, they will meet with a second and more impassable defence, which is set round with a sort of Spikes, which all meet in a point or centre, something like the contrivance which is common to be seen in some sort of Mouse-traps, which easily admits the Animal one way, but wholly forbids and opposes its return.

This will be better understood by viewing the Figure at *(t,)* which is half of a Case, supposing the Case to be disected length-ways.

The SMALL EGGER.

THE Hen Egger lays her Eggs toward the latter end of March, which she disposes round a Twig, these are covered with a downy or woolly Substance, which she does thus; viz: every Egg she lays, which is covered with a gummy or glewy Moisture, she rubs her Tail all over it, and the Down with which her Tail abounds, adheres to the Egg, and are so pulled from her Tail. I do not speak from imagination but from experience having oft-times seen them laying their Eggs. These Eggs are hatched in April, when the young Caterpillars begin to spin a Web, which they extend larger as they increase in size, covering it over with a fresh Spinning, every Skin they shift, that like an Onion it consists of several Coats. On this Web they all lie in fair Weather; but when it rains or the Wind blows too hard, they all run in for shelter.

BESIDE this Web, they extend others to the several branches of the Food, which serve as ladders from and to their Web, which they never desert, without the Bush which is generally very low, is stript bare of Food. When full fed, as seen at *(k)* they retire down the Stalk, and finding a convenient place, spin themselves up in Cream coloured Cases, like that seen at *(l;)* wherein they change to their Chrysalides which are short and thick and of a light Nut-brown colour, as seen at *(m.)* They lie in that state during the Winter, and the Moths appear about the middle of March. It is a very difficult Moth to bred from the Caterpillar, and should the young Aurelian take home any Caterpillar with intent to breed the Moth, and not take the Web with him, they will all certainly die; notwithstanding they possibly may feed with him till they be full fed. The upperside, of the Moth is seen at *(n,)* and the under at *(o.)* They are seldom taken in the Moth state.

The YELLOW TAIL.

THE Eggs which are of a pale green colour are laid by the Hen about the beginning of July, these are covered over with a woolly Substance as the former, and produce their Caterpillars about the beginning of August; which feed on Blackthorn, Whitethorn, Oak, Willow, and most sort of fruit Trees. At the approach of Winter they spin themselves little Cases, in which they remain during the Winter, in Spring they come forth, feed again and become full fed, as at *(p,)* the beginning of June, when they change to the Chrysalides, as seen at *(q,)* within a spinning. And the Moths appear in about three Weeks after; I have described the Hen at *(r,)* in the odd position in which they always sit, and the Cock at *(s;)* which only differs from the Hen in the Horns, the Cocks being Comb-like, and broad. The Moths are very common and easily taken.

Plate XXV

The beautiful colours of the Emperor Moth, Pavonia pavonia *and the eye-like markings on the wings rival the Peacock Butterfly for the most handsome of Britain's native lepidoptera. It is not by any means rare in the British Isles and in some places may be found commonly even in the London area, where it was discovered by Harris. The unique cocoon which allows the moth easily to emerge in due course, but prevents the ingress of any insect, is well described, but here again the assumption that differently coloured caterpillars are either male or female is mistakenly made. The food plants of the caterpillar are varied, including heather, bramble and sallow, as well as those upon which Harris found it.*

The conspicuous web described in this account of the Small Egger, Eriogaster lanestris *must have been a familiar sight on hedges of sloe or hawthorn in the last century, but its numbers have so decreased that now it is found only in widely scattered localities in the southern half of England and northern Ireland. Formerly it could be found over the whole of England as well as parts of Scotland and Ireland. Modern farming methods and pollution have been the cause of this serious decline. The moth that Harris has depicted is a female, which is considerably larger than the cock, and is furnished with a large anal tuft, the purpose of which is explained in his account. Several species of moth remain as pupae for more than one year, but this one has been known to spend up to seven years in that stage, awaiting to emerge when the time is right. Surely a record for any English lepidopteron.*

The scientific name of the Yellow Tail is Euproctis similis. *The male of this white moth is somewhat smaller than the female, and usually has a small dark mark on the forewings. It can be found, rarely, in southern Scotland and, more frequently, throughout England.*

M.ʳ Harris ad Vivum

To the Right Honourable _____ the Countess of Berkeley

This Plate is humbly dedicated by her Ladyship's   most Obedient Humble Servant

DIEU AVEC NOUS

Moses Harris.

M? Harris ad Vivum

To the Rt. Hon.ble Lady Spencer this Plate is most humbly Dedicated by her Ladyship's most obliged humble Servant. Moses Harris.

DIEU DEFEND LE DROIT

Plate XXVI

The GREEN FLY.

The BRAMBLE. *Rubus major fructu Nigro.* J.B.

THE Caterpillar is extremely scarce and difficult to find. They feed on the Blossoms and Buds of the Black-berry or Bramble, hiding themselves under the smaller green Leaves: the best time to search for them, is about the latter end of June or beginning of July, when some of them are pretty large; for about the latter end of July, they are full fed, when they appear as represented at *(a)* and *(b,)* which last is drawn as fell from the leaves on his Back intended to shew the flatness of his Belly, they now retire and hide themselves in convenient places in order for their change where they fasten themselves up by the Tail, and round the middle, with small and tender Threads, in the manner as I have before related of the Hairstreak. Thus prepared, they change to their Chrysalides which appear like that at *(c;)* they remain in this state during the Winter, and the Flies appear in the middle of April. They fly in Woods, and may oftentimes be taken in plenty. The Male and Female are much alike, although they differ something in size; the Hen being rather larger and the small light spots in the underside, the under-wings are rather more conspicuous. It is almost in vain to look for the Caterpillar except on such Bramble Bushes, where you have seen the Flies playing in plenty, which you sometimes may in Woods, then return at the Time when the Caterpillars are pretty near full fed, and while some one holds the beating Net, beat the Branches and Heads of Blossoms; by this means, and searching the Buds, you may be almost certain of finding the Caterpillar. The Male is seen flying at *(c)* shewing the upper Side of his Wings, which are of a deep brown Colour: the Female is described at *(d)* with her Wings erect, to shew the under Side. It is remarkable in this Fly, that when pursued it will settle on the Branches of some small leav'd Bush, such as the Blackthorn, with its Wings shut; and as it then appears all green, it is so like the Leaf that I have very often sought for it in vain, altho' I thought, at the same time, that I had taken exact notice of the place where it settled.

The DARK GREEN FRITILLARIA.

THE first Appearance of this Fly is about the Beginning of June. It is extremely timorous and swift of Wing; and should you, as it flies by you, strike at it with your Nets and miss it, it is in vain to pursue it; for, being frighted, it is wild, and will not settle till it be quite out of your Sight. The best Time to take them is toward the Evening, when they rest, and feed on the yellow Flowers of the Hawkweed, and the Blossoms of the Bramble, in the Recesses of Woods, where it is open to receive the Sun-beams, and a Shelter from the Wind; then, if you tread softly, you may come very near them. This Fly was never bred from the Caterpillar that ever I could hear, but a Gentleman of my acquaintance informed me that he once got one of the Caterpillars which he found on the Ground in a Wood, about the Beginning of May; that it was of the spiked or prickly Kind, and appeared in all other respects so like the Admirable, that he was deceived. It changed into Chrysalis, hanging by the Tail, in two or three Days after, but died in that State, just before the Time of its Appearance in the Fly State, which he said he knew by the Spots on the Wings, appearing through the Chrysalis; finding it did not come out, after waiting some Days, he broke it open, and discovered it to be the Dark-green Fritillaria. I have shewn the Female flying at *(c)* displaying the upper Side of her Wings: the Male is seen at *(p)* as settled with his Wings erect to shew the under Side, which of Male and Female are finely spotted with silver; the Female lays her Eggs about the Middle of July, and the Caterpillars continue in that State during the Winter, but, what they feed on has never yet been discovered.

The BLACK OVAL WATER BEETLE

ABOUT the latter End of June and the greatest Part of July the Caterpillars of this Beetle may be found of different Sizes. When full fed they appear like that at *(e)* of a deep Olive, or dirty green Colour on the Back, the Belly is something lighter. Their Food is commonly the Caterpillars of other Water-insects; indeed they are so rapacious that they will destroy Insects much larger than themselves, or any thing that comes in their way, as Flies, small Moths, that by accident fall in the Water, and the small Worms, which are by Fishermen called Blood-worms, they are remarkably fond of. When ready for their Change, they return to the Bank, where, in some of the holes of the broken earth, they get in, and make themselves a hollow Cave, wherein they change to the Nympha, as seen at *(f)* the Beetles appear toward the latter End of the Summer, and continue during the Winter in that State I have described the Beetle crawling at *(g)* to shew its Back; at *(h)* it is shewn lying on its Back, shewing its Legs, and the Formation of its Belly. Again at *(i)* I have shewn it flying, whereby the thin membraneous Wings are discovered, as well as the Formation of the naked Back, which are concealed when the Shell-wings are shut.

The FLAT BLUE-TAIL'D LIBELLA.

THE Caterpillars are commonly found in foul Ditches and stagnated Waters, in April and May, of different Sizes. These likewise feed on other Water-insects, and become full-fed in the Month of June, and appear as at *(k)*; they then crawl up out of the Water by the help of a Stone, Stick, or Piece of Grass, where, holding fast by their Legs, they presently burst on the back, and the Libella draws itself gently out, as at *(l)*, it then holds by the Legs, with its Wings hanging downwards, where they gradually stretch and dry, and are quickly fit for flight. There are two Sorts of this Libella, the one with blue Tails, the other with brown: these are allowed to be Male and Female; that with the blue Tail is supposed to be the Female. The Libella at *(n)* is of a different Class of Libellas, though by some it has been mistaken for the Male of the Blue-tail'd; and I had nearly fallen into that Error myself; for I happened to be out one Afternoon and perceived a Blue-tail'd Libella and another of the above-mentioned Sort playing together over a Pond of Water; they wantoned about, and pursued each other a long while, often settling one atop of the other, as though going to copulate, so that I really concluded it to be the Male of the Blue-tail'd Sort: But I was soon convinced of my Error. The Blue-tail'd Libella is described in the Plate at *(m)*. The Caterpillars of almost all the Libellas are to be found in such Ponds as I have described, in April or May. The Method used in taking them is, with what the Fishermen call a Landing-net, fixed to the End of a long Stick, with which the Mud and Weeds are taken up from the Bottom of Ponds, in which you are to search for the Caterpillars. The Ponds in and about Woods are generally the best.

This pretty little insect, the Green Hairstreak, Callophrys rubi, called by Harris the Green Fly, was first described by Merrett in 1666. Although the impression given by Harris is that the caterpillar feeds only on bramble (and indeed in his English Lepidoptera of 1775 it is named as the 'Bramble or green fly'), in fact it has a very wide range of food plants. These include gorse, broom, trefoil, dogwood and buckthorn, in addition to the bramble from which its scientific name derives. It is widely distributed throughout the British Isles, 'about London' being included by Petiver in 1702. It is said to be Britain's most common Hairstreak.

The food plant of the Dark Green Fritillary, Argynnis aglaja is the heath dog-violet. This butterfly, which is by no means scarce, occurs over most of Britain in areas of open country. It usually flies in July and August, and the larvae hibernate as soon as they are hatched, before feeding. First described by Moffet in 1633, it was known also as the Silver Spotted Fritillary to aurelians of the last century.

There are several species of water beetles which could have been found by Harris in ponds in the London area, and this appears to be one of those in the genus Ilybius, all of which have a general resemblance to each other. Probably the painting is of I. ater, a black species which likes stagnant water, or another smaller black species I. obscurus, with which it was frequently confused by early authors.

The specific Linnean name which Harris gave to the Flat Blue-tail'd Libella, Libellula depressa was correct. There can be no doubt about this for it is the sole species of British dragon-fly so broad in the body; it is now called the Broad-bodied Libellula. Commonest locally in the southern counties of England, its presence and numbers may also be accounted for by immigration. It is the adult male which Harris painted for the female has no blue colouring; in her case the blue is replaced by brown. Harris rightly dismissed the idea that the other species he depicted was also depressa for the female is even more ample in width than the male. The representation of this dragon-fly's wanton playmate corresponds closely to a species which is now very rare, although it does occur in a few scattered localities in England, but not in the London area. This is the Scarce Libellula, Libellula fulva and the fact that it was recorded from the marshes around London in the first half of the nineteenth century adds to the probability that this was the insect watched by Harris and whose identity so confused his contemporaries.

The RUBY TYGER.
The SWEET-SCENTED PEA.

THE Moth is generally bred in the Month of *June*; and the Hen, after copulation, flies in search of a convenient and safe Place to deposite her Eggs; which she disposes of in great Regularity and exact Order, as at *(h)*: these are green and glossy. The young Caterpillars, when they proceed from the Eggs, wherein they lay about fourteen Days, are of a light Olive Colour, covered with light tender Hair, which appears darker in every Skin. In about the third Skin they cease feeding, and conceal themselves during the Winter. Early in the Spring they begin to feed again, appearing in their last Skin about the Middle of *May*; some much sooner, others later; for I have found the Caterpillar, in the Middle of *June*, feeding on *Ragwort* and likewise on *Hounds-tongue*; for, like others of this Class, they will refuse hardly any Thing.

When full fed, as at *(i)*, they prepare for their Change, by spining themselves up in a slight Web, as at *(k)*, wherein they change to a black Chrysalis, as shown out of the Web at *(l)*. They lie in this State about twenty-eight Days, and then the Moths appear; one of which is exactly described in the Plate at *(m)*. This Moth is not very plentifully found; yet an Aurelian of my Acquaintance has a Hen which flew into his Window in London, and layed near an hundred Eggs, which produced Caterpillars. These were fed on Lettice; but, being sheltered by the House from the Intemperance of the Weather, they fed and shifted their Skins very fast; and, even after the cold Season was very far advanced, so that most of them were in their last Skin, and too near their Change to undergo the Severities of the Winter: and it is an Observation we in general make, that, when any Caterpillars, which are to continue during the Winter in that State, are too forward, we have little or no Expectation of their producing any Flies; for they commonly all die in the ensuing Spring; as was the Case here.

The WALL.

THE first Appearance of this Fly is about the Middle of *May*; and, towards the End of that Month, they begin to lay their Eggs on the Grass; which are, without Doubt, fixed fast thereon by a tennacious Matter. The Caterpillars feed thereon, and become full fed about the Middle of *July*; and appear as represented in the Plate at *(a)*. They then hang themselves up by the Tail, as at *(b)*; where, after about two Days, they change into a short, thick, green Chrysalis, of a remarkable Form, as described in the Plate at *(d)*; which is a profile Draught of the Chrysalis. The Chrysalis at *(e)* shews its back Parts, that the true Shape of this Chrysalis might be the better comprehended. The Fly appears about the beginning of *August*. The Caterpillars proceeding from these, continue, during the Winter, in that State; change to Chrysalides about the beginning of *May*; and the Flies appear about the Middle of the same Month, as above. The Male and Female differ pretty much in the Markings of their Wings, as well as in other Respects, as to Colour and Size; as may be seen by comparing them in the Plate. The upper Side of the Female is seen at *(f)*, and the under Side at *(h)*. The upper Side of the Male is seen at *(g)*, where it is represented as fixed in a Chip Box with a Pin. This Fly is very common in Fields, and by Road-sides. It delights to fly along very low in dry Ditches, seldom straying from the Bank, or Field, where it was bred; but, when it comes to the End of the Bank, will return back again, frequently settling against the Bank, or perhaps against the Side of a Wall; and is, for this Reason, called THE WALL FLIE.

The WHITE SPOT.

OF this Moth little or nothing is known, but that it is taken in Woods in the Month of *May*. I have often taken it in *Cain Wood*, which is between *Highgate* and *Hampstead*; but they are now taken in the greatest Plenty at *Tottenham Wood*, in the Cutt. There is an Alley in *Oak-of-Honour Wood*, which is known by the Name of *White Spot Walk*, which leads from the *White Admirable Walk* to the *Long Gallery*; but whether the *White Spots* ever flew there in Plenty, they best know who first agreed to call it by that Name. I have given a Draught of the Moth in this Plate at *(p)*.

The DUKE of BURGUNDY, FRITTILLARIA,

COMMONLY called *the Burgundy*, is one of the four Fritillarias which want the silver Spots, and is the least of them all. They always fly in Woods, and not very high above the Grass. Their most plentiful Time of Flight is about the Middle of *May*. They are very nimble, yet I cannot say they are difficult to take. I never saw them fly in Plenty any where but at *Comb-Wood*, though I have known them to have been taken in great Plenty by others in several other Places.

As to the Caterpillars and Chrysalides, they have never yet been discovered by any us.

I have described the upper Side at *(n)*, and the under at *(o)*, where it is shewn as sitting with its Wings close.

Plate XXVII

Sometimes seen flying in the sunshine in May or June, this little moth, the Ruby Tiger, Phragmatobia fuliginosa, *has emerged from a pupa formed earlier in the year from an overwintered larva. Despite the author's experience of hibernating caterpillars, it has been observed that this one does not cease feeding until it has reached the last stage before pupating, and in this instar it is able to withstand extreme climatic conditions.*

The Wall Butterfly, Lasiommata megera *has a fondness for settling on walls or banks where it can spread its wings in the sun. Ray delightfully called it the 'Golden Marbled Butterfly, with black eyes'. It was doubtless common in or near London when Harris collected, for to Petiver it was known as the 'London Eye'.*

Anania funebris, *the life history of which was so much a mystery to Harris, is one of a family of moths known as Pyralids. It is a little slender-bodied insect, whose yellowish caterpillar feeds on the flowers of the golden-rod until the autumn, when it spins a cocoon in which it overwinters until its pupation in the spring. The attractive names given by the enthusiastic aurelians to their favourite hunting grounds sound a delightfully rural note — names which, alas, would now be appropriate only many miles from the metropolis.*

This little butterfly, the Duke of Burgundy Fritillary, Hamearis lucina *is the smallest, of the Fritillaries — although strictly speaking it is not a true Fritillary. The caterpillar, which was unknown to the aurelians, seems to feed exclusively on cowslip or primrose. It is of a pale brownish colour and changes to a chrysalis in August, in which state it overwinters until emerging in May. It is a rather local woodland species, found mainly in the south of England and the midlands, and was noted by Petiver in his* Papilionum Britanniae (1717) *as occurring 'in several woods round London'. It was described there under the name of 'Vernon's small Fritillary' after Ray, in* Historia Insectorum, *commented that it 'was first observed by Mr Vernon, about Cambridge'. No one yet seems to have discovered why its attractive common English name was originally applied. Perhaps, as it occurs locally over most of the Continent, its European name was simply adopted there.*

M⁙ Harris ad Vivum.

To Her Grace the Dutchess of Norfolk
This Plate is humbly dedicated by Her Grace's most devoted Servant.
Moses Harris.

SOLA VIRTUS INVICTA

To Lady Ecklin

her Ladyships most Obedient

This Plate is humbly Dedicated by

Serv.t Moses Harris

Plate XXVIII

From the aristocratic name of the last insect described by Harris we now come to the least attractive one he used; the 'Dishclout, or Greasy Fritillaria'. First described by Ray as a British species, it was later (in 1717) called by Petiver, 'Dandridge's midling Black Fritillary', and is now known as the Marsh Fritillary, Eurodryas aurinia — not a very exciting name, but an improvement on Harris's choice. Dishclout, forsooth! It is widespread over most of Britain, but only in scattered local colonies. In some areas, such as East Anglia, it is very rare or extinct; certain it is that Harris would never find today this pretty butterfly at Neasden, where he found it in 'great Plenty'!

Although his supposition that the caterpillar of this butterfly fed on violet was correct, his belief that it spent the winter in hibernation was mistaken. The High Brown Fritillary, Argynnis adippe lays its eggs in July or August and these do not hatch until the warm days of the following April. It is a declining woodland species found locally over parts of England and Wales. Harris copied Jacob L'Admiral's painting of the caterpillar and chrysalis, which were unknown to him. However, some considerable doubt must be thrown on these two representations, which are marked 'a' and 'b' on the plate but with no text reference. The lack of colour on the plate suggests that either Harris or his colourist omitted it in error.

It was not confirmed until 1923 that the moth illustrated here was distinct from another very similar species. The two moths are the Treble-bar, Aplocera plagiata and the Lesser Treble-bar, A. efformata. They have similar characteristics, the caterpillars feeding on various species of St John's-wort, and both overwintering as pupae. Very slight differences in the wing pattern are the main distinguishing features, and these indicate that Harris's moth is probably the Lesser Treble-bar.

A bright little spring moth is the Speckled Yellow, Pseudopanthera macularia and the sight of them fluttering about the furze bush in profusion must have gladdened the heart of Moses Harris. His labour in beating the bush for its caterpillars was in vain for they have a different taste and feed on wood-sage, dead-nettle and other plants.

Pyrausta purpuralis is a colourful moth, as can be seen from the plate, aptly named by Harris the Crimson and Gold. It is of the same family of Pyralids as the White Spot of the last plate, but this one is much more brightly arrayed.

The DISHCLOUT, or GREASEY FRITILLARIA.

The Devil's Bit, *Scabiosa Radice Succisa*, Ray's Syn.

THE best Time to find the Caterpillars of this Fly is from about the Middle of *April* to the Beginning of *May*, being then almost all in their last Skin. They may be found in Plenty the latter End of *March*, but then you can have no Food for them, it not being then come out of the Ground. They are likewise so troublesome to breed, (for they never will eat except the Sun be upon them), that it is better to stay till they be nearer their Change. They are generally found on the Side of a Hill that rises with an easy ascent and fronts the East; by which they have the Sun most powerful in a Morning, and avoid his too scorching Heat in the Afternoon. It was said that they fed on Plantain and Grass, but I found that to be a Mistake, having often endeavoured to feed them with both, but my Endeavours were always fruitless. At length I was determined to come at the Truth; and accordingly, on the Eighteenth of *April*, 1760, I went to *Neesdon*, near *Welsdon*, about seven Miles from *London*, where I was informed they were in great Plenty, as indeed I found them to be. Here I took great Pains to watch their Actions for full three Hours. I paid them several Visits a few Days afterwards, that I might be capable of giving a satisfactory Account of them. Their Food is the *Devil's Bit*, which, at that Time of the Year, hardly appears above Ground. They feed on the opening Leaves as fast as they come up; which is the Reason why those who found the Caterpillars could never see the Food. When the Sun happens to be shut in by the Clouds, they immediately stand still, and, though eating very greedily, they will suddenly cease; but, on the Return of the Sun-beams, they run nimbly over the Tops of the Grass, and descend into every Vacancy of the Grass they can find in search of their Food. Nor did I ever find above two at one Root, although the Field appeared to be covered with the Caterpillars. When in their last Skin they appear, as at *(e)*, very black and thickly set with branched Spikes; and their Backs and Sides are powdered with white specks. The Preparation they make for the Preservation of their Chrysalides is much to be admired. When one is ready for his Transformation, his first Business is to draw several Pieces, or Blades, of Grass across each other toward the Top. These he fastens together with his Webb; and then beneath the centre, where the Blades of Grass intersect each other, he hangs himself pendulous by the Tail, and changes to the Chrysalis, which is justly described at *(f)*. This Method they have of providing for their Safety while in the Chrysalis State, is a strong Proof of the amazing Instinct of these little Creatures. They are not only hereby hid from the Sight of Birds, but defended from the Damage they might otherways sustain in boisterous and windy Weather; for, as the Grass is drawn from every Side, let the Wind blow which Way it will, one or more of the Pieces of the Grass immediately acts in the Manner of a stay. The Flie appears from the Beginning to the Middle of *May*. The Female is described, shewing her upper Side, at *(g)*, and her under Side at *(h)*. The Male is seen at *(i)*, shewing his upper Side. The Eggs of this Flie I never saw; but, as I know it is the Nature of this Class to fasten their Eggs to the Food, I should suppose these so to do. As to the Caterpillars, I am pretty certain they continue all the Winter in that State, for I have found them early in *March* (which is before their Food appears) in about their third Skin, where they have lain all together on a dead and withered Leaf under a thin Web. I have opened the Web and found them numbed, and as it were lifeless; therefore it is not in the least to be doubted but they had remained so from the beginning of the Winter. It is remarkable in this Insect that neither Flie nor Caterpillar will stray from the Field in which they were bred; and, though I have seen some thousands in the Field, yet I never could find one in the Meadows adjoining. They fly very low over the Tops of the Grass. The Reason why this Flie is called THE GREASEY FRITILLARIA, is, because the under Side of the upper Wing always appears greasey.

The HIGH BROWN FRITILLARIA.

THIS Fly may be caught in plenty about the beginning of June, or at the same time and place as the Darkgreens; they fly with great rapidity, therefore, taken with some difficulty. The Caterpillar and Chrysalis, has never yet been discovered in England: I have taken a great deal of pains to find it, but all to no purpose. I should suppose, according to Mr. Admiral, from whose inimitable work, I took the Caterpillar and Chrysalis in the plate, they should feed on the violet; but I have often searched for it without success, the time to seek it is about the middle of May; for these likewise lay, during the winter, in the Caterpillar state. The upper-side of the female is seen at *(c)*, and her under-side is seen at *(d)*, where it is shewn as settled with its wings erect, on the flower of the Heart's Ease, extracting with its proboscis, the liquid honey: the male is like the female, but much less; they delight to fly in woods and settle on the blossoms of the bramble.

The TREBLE BARRS.

THIS pretty Moth is generally taken flying, in and about woods, any day towards the latter end of May, in the morning, just after sun-rise. The Caterpillar has not yet been discovered, and, altho' the Moth is now taken in some plenty, yet, some years ago it was esteemed a great scarcity. I have described it in the plate at *(k)*, shewing the upper-side of his wings; the under-side is of a very pale, brownish colour, without any markings; I therefore thought it needless to shew it.

SPECKLED YELLOW.

ABOUT the middle of May, is the time to go in search of this Moth; they are always found in woods, and in the day time seem remarkably sociable, for the most part haunting some particular place in the woods; altho' it is confest, they may be taken or met with, singly, and that in almost any part of the wood at the same time; yet I observe, there is always a favourite place or spot, in every wood where they breed, which they most affect, and are there seen in great plenty; sometimes I have seen great numbers hovering about the furz-bush, that at a distance (they being nearly the same colour with the yellow flowers of the furz) it appears, as if the leaves of the flowers were blown about by the wind. These Moths appearing to me so particularly fond of this bush, made me almost believe the Caterpillar fed on it; but though I have beat it many times, I never could find that any Caterpillars feed thereon. I have shewn the Moth flying, at *(m)*, shewing the upper-side; the under-side is marked as the upper, excepting that on the under-side, the upper wings are marked strongest, and contrarily, the under wings are strongest on the upper side.

CRIMSON and GOLD.

THIS richly coloured Moth, is generally taken by beating the nettles, which is therefore supposed to be the food of the Caterpillar; but it has not, however, yet been found. I have shewn it flying at *(l)*, displaying its upper-side, the under-side is nearly marked the same as the upper. The best time to take this Moth in its beauty, is about the middle of May.

The GREAT-EGGER.
The WHITE-THORN.

I HAVE before, in the Twenty-fifth Plate, given the history of the Small-Egger, I shall now proceed to give the best account I can of the Great-Egger. The Hen Moth lays her eggs about the beginning of June, she does not fix them to the food, but drops them loosly to the ground; they are of an oval form, of a light brown colour, prettily mottled; they continue in the egg state about a month, but I have known them lay five weeks before they have produced Caterpillars; but those Caterpillars have proved as healthy and strong, as those which have not laid so long in the egg. The young Caterpillars feed on the White-thorn, during the remaining part of the summer, but when the leaves are strip'd off by the approach of frosty weather, they continue to feed on Laureltine, and other ever-greens, growing but slowly during the winter. When full fed, as at *(a)*, which is about the middle of May, they are about four inches long, covered with brown hair, across each joint is a very black stripe, which stripe is naked, tho' appearing like velvet: along each side it hath many irregular white marks which lay a little oblique; it has likewise, two little tufts or tussocks, one on each side the head, appearing like ears, their motion is indifferently slow. When fit for their change, they spin themselves, each in a long, oval, brown case as a *(b)*, this is thin, and of a tolerable strength, or of the substance of pretty thick paper. The Chrysalis is shewn at *(c)*, as taken out of the Case. In this Case they continue about one Month, when the Moths appear. The Cock shews his upper Side at *(e)*, and the under is discovered at *(f)* where he is seen as hanging by a Sprig, to dry and stretch his Wings, supposed to be just come from the Chrysalis; the Hen shews her upper Side at *(d)*; she is larger, and much paler than the Cock. The Aurelians take this Moth by Sembling; their Manner is, to go out with a live Hen in a Box *(a)*, which is covered down with Gauze or Crape; when they are come to the appointed Place, where they are pretty certain there is a Brood, they set the Box on the Ground, and stand ready with their Nets; the Cocks will quickly come and attempt to get at the Hen. I have known great Numbers taken in one Hour's Time; and it may be depended on, that, if any one goes with a Hen, in almost any Place of the Country, they will not fail of Success; not only the Egger and Vapourers, but any Moth may be taken by Sembling; but the above-described Method is chiefly used to the last-mentioned, As to the other large Kind of Moths, such as the Private Poplar, Emperor, Lime, &c. &c. the usual Method is, to tie the Hen to a Tree, Bush, &c. lightly tied or fastened round the Body with a Piece of sewing Thread, and there to be left all Night, and in the Morning, when you return, you will almost be certain to find Madam accompanied by her Spark, who will not desert his Mistress, though her Favours be ever so easily obtained.

The YELLOW, or BRIMSTONE MOTH.

ITS Caterpillar feeds on White-thorn, is of the Luper Kind, and remarkable for a Protuberance in the Middle of its Back, as is seen at *(g)* and *(h)*; this Protuberance is sometimes divided, and appears like two. It changes into a Chrysalis in the Month of September, and the Moth appears in the Middle; or towards the latter End of May. Its Chrysalis is shewn at *(i)*, the upper Side the Moth is described at *(k)*, and the under at *(l)*, where it is seen hanging to dry its Wings, as just come out of the Chrysalis. It is to be taken flying in an Evening, in Lanes which are hedged with Whitethorn; it is easily taken, and is commonly found in Company with the Silver Ground.

CLOUDED YELLOW.

THIS beautiful Fly is taken in Meadows, in the Month of August; they appear fond of settling on the Yellow Lupins and Thistle. They have been taken flying, in Plenty, on Epping Forrest; but as they seldom haunt one Place for many successive Seasons, I can't venture to mention it as a Place where they are to be found. Where there is a Brood, the Times of the Day to find them are at Eight in the Morning and Four in the Afternoon; but never in the Middle of the Day, when they conceal themselves to rest. They fly very fast, therefore not easily taken; the Male, in particular, flies exceeding swift. I have shewn the upper Side of the Female at *(m)*, and that of the Male at *(n)*; the under Side of the Male is shewn at *(o)*. Their Caterpillars were never found.

CHINA MARK LIKENESS.

THE Caterpillar of this Moth is of a bright Yellow, and full of black Spots, as is shewn at *(p)*; they are commonly found spun up in the Nettles, as is shewn at *(q)*. In the Beginning of May they change into Chrysalis, which is seen at *(r)*. About the Middle and the latter End of that Month the Moths appear. The Moth is described in the Plate at *(s)*.

The WOOD WHITE.

THIS is the least of all the White Flies, and is always taken flying in or near Woods; 'tis remarkable for the Smallness of its Body, it being accounted the least of the Butterfly Kind, though not the shortest; I likewise observe they seldom are seen to settle. They fly twice a Year, or at the same Times with the large and small Garden Whites, viz. May and August. The Caterpillar is utterly unknown, and little Expectation there is to believe it ever will, as it must be very small, according to the Fly, which is seldom seen but on its Passage. It is described in the Plate at *(t)* and *(u)*.

Plate XXIX

The males of the Oak Eggar, Lasiocampa quercus, called here the Great-Egger, are probably known better than any other moth for the ease by which they may be captured when using the 'assembling' method. It is the male which is more frequently seen, for it flies in the sunshine very speedily and erratically and, as it is a large heavy moth, very conspicuously. It is in this wild careering flight that he is diligently searching for a female, hoping to pick up the scent by his heavily pectinated antennae. The recognition that the male moths of many species may be attracted in this fashion reaches back long before Moses Harris, for John Ray recorded that in 1693 two specimens of the Peppered Moth were taken by his wife one night in a room in which a female had just emerged from a chrysalis.

The Brimstone Moth, Opisthograptis luteolata is common throughout the British Isles and, together with its companion (called by Harris 'the Silver Ground'), it frequents lanes and hedgerows. Not described or illustrated in The Aurelian, *the latter moth is almost certainly the Silver-ground Carpet, Xanthorhoe montanata.*

It is not surprising that Harris was unable to find the larva of this fine butterfly, the Clouded Yellow, Colias croceus, as it is an immigrant, unable to overwinter in Britain in any stage. The numbers arriving in England fluctuate considerably, but in most years specimens of this butterfly are recorded from several counties — sometimes in large numbers, as in 1947 and 1983.

Udea olivalis is the scientific name of the little Pyralid moth which appears to be the one Harris calls the China Mark Likeness. It is found plentifully over the British Isles except in the north of Scotland, hiding in the daytime in hedgerows.

After commenting on the butterfly called the Wood White, Leptidea sinapis, Moses Harris seemed to be in despair that the caterpillar would ever be found; he obviously underestimated the enthusiasm with which following generations would pursue his favourite study! The slow flight and slenderness of the body render the Wood White a most delicate butterfly, and the white wings add to the impression of a summer snowflake. It is a very local insect, found in southern and western counties of England and in Ireland, delighting in woodland rides. Petiver found it in Hampstead, but its haunts near London have long disappeared. It is interesting in that there are two slightly different broods in the year, the spring one being rather darker than the summer brood.

PL·XXIX

To the Hon.ᵇˡᵉ   Richard Bateman
This Plate is humbly Dedicated by his most   Obedient Servant Mofes Harris.

To the Honble John Ward

This Plate is humbly Dedicated by his most Obedient Servt Moses Harris

Plate XXX

CRABTREE IN BLOOM.
The FIGURE EIGHT.

IT is plainly seen why this Moth is called by this Name, as the Figure Eight is so conspicuous on the upper Wings. The Caterpillar, which is seen full fed at *(a)* feeds on Blackthorn, Whitethorn, and Crabtree, and is taken by beating in May and June; changes to Chrysalis in a pretty hard Case, as at *(b)*; and the Moth appears in the Month of August. I have shewn the Female flying at *(c)*, and the Male, as settled on the Stone, at *(d)*; 'tis known from the Hen by the Broadness of his Horns, which are ramified, while those of the Hen are like small Thread.

BROWN PLUMED.

I Found one of these Caterpillars on the 6th of August, feeding on the common thistle; it appeared of a flesh colour, but on a nearer view finely freckled with variety of different tints: the whole Caterpillar was flat, and the belly seem'd to adhere to the leaf somewhat like the snail, for there was no appearance of any legs; it was broadest in the middle, and from thence grew taper toward the head and tail, which was very small; and I believe the whole Caterpillar might be about three quarters of an inch long. I did not disturb it by taking it off from the leaf, but brought it home, together with as much of the plant as I thought convenient, and set it in a bottle of water in one of my cages; it seem'd to feed pretty hearty till the 11th, when it ceased feeding; and the next day it fixed itself up by the tail against a bit of small dry stalk put by chance in the bottle together with the thistle; the head part of the Chrysalis was upward: a little surpriz'd at the oddness of the position in which it had fixed itself (which is very rare in the moth kind) I ventur'd to touch it, to see if it was secur'd by a thread round the middle, which those that fasten themselves erect in this manner commonly are, when I was agreeably surpriz'd at its giving itself a sudden spring backward, its head then hanging down, as is shown at *(k)*, the motion of the head being described by a dotted line: he did not remain above two seconds of time in this position; and then, with as swift a motion as before, sprung upward and reassumed its former posture. I waited with much impatience till the 22nd of August, when it produced the brown Plumed Moth shewn at *(l)*. It has been supposed, and many have insisted on this being the male to the white Plumed Moth; but their being produced at two different times of the year, and their respective haunts so greatly disagreeing, these for the most part being found in woods, and the other on nettles by ditch sides, I am tempted to conclude them of different species.

SCARCE SILVER LINES.

THE Caterpillars are taken by beating the oak trees which grow in hedges the latter end of May, 'tis all over of a fine green, and is remarkable for a riseing on the back towards the head. It becomes full fed, as described at *(e)*, about the beginning of June, when it prepares itself for its transformation by securing itself in a pretty strong case, the form of which is described at *(f)*; this is made fast to the branch or body of the tree. Within this case it changes to a white Chrysalis, which is remarkable for a broad black mark down the back, as seen at *(g)*. The Moth appears the latter end of June, and is described in the plate at *(h)*. This is call'd the Scarce Silver Lines, on account of its similitude to that in the tenth plate, but the lines in this Moth, are not of that silver or glossy white, as in that; but of a bright yellow, or straw colour.

The WHITE ADMIRABLE.

IN all my researches in the Insect world, I have not been able to discover the Caterpillar of this excellent Fly. I have watched the females several times in woods, thinking, to find them laying their eggs; I have likewise beat every tree and shrub I could think on, about a month before their time of flight, but to none effect; so all that I can inform my reader is, that the Fly may be taken in woods, where they are found in plenty the latter end of June, and beginning of July; they fly very rapidly, often skiming like a swallow, and are fond of settling on the leaves of the oak; sometimes they settle on the ground, in the shady paths of the woods: they are very timorous, and when persued, with wonderful swiftness dart over the tops of the highest trees, or settle on the topmost branches, where they will be sure to tire your patience ere they will remove. By a gentlemen lately arrived from some part of Germany, I have received an account, that this Fly lays her eggs on the very tops of the highest poplars; where they are hatched, and remain during the winter, in the Caterpillar state, securing themselves by their web or spinnings; that in the spring they feed again, when awakened by the warmth of the sun from their dorment state, and become full fed the beginning of June; change to the Chrysalis, hanging by the tail; and the Fly is produced the end of June; this account certainly carries with it a great appearance of truth, more especially as it is similar to the history of several of our English insects, in particular the Purple Emperor; and indeed I did ever consider the White Admirable of that class; I have described the upper side at *(m)*, and the under at *(n)*.

CHIMNEY SWEEPER.

THIS sable-coloured Moth is never seen but in woods, and their time of flight is about the beginning of June; they fly very low, and are easily taken; they do, for the most part, frequent some particular spot in the wood, where many of them may be seen together, and 'tis but seldom that one is seen alone; 'tis described in the plate at *(o)*. I thought it needless to describe the underside, it being semular to the upper.

RED ARCHES.

THIS Moth is generally taken by beating the boughs of the oak, in woods, the latter end of June, and beginning of July; the most likely oaks to answer the purposes, are such as are used to be cut away every few years for faggoting. For this use we commonly provide a piece of cloth, about four feet square, with sockets on two of the sides, in the manner of the batfolding netts; through these sockets, two sticks are put, of a moderate length, so as conveniently to hold the cloth extended under those short trees, while another beats the boughs. The Caterpillar and Chrysalis remain unknown. When it falls in the sheet it seems to all appearance dead, and will endure handling pretty roughly before it will stir. It is described in the plate at *(p)*, shewing the upper side.

The SNOUT-EGGER LIKENESS.

THIS Caterpillar may be found feeding on nettles the beginning of May, at which time it is very full fed, and appears as at *(f)*; it is of the Quarter-Luper kind, and about an inch or better in length; its colour is green, with two white lines on each side the body, which run from head to tail; it hath some few hairs; when ready for its transformation it spins a slight web on the surface of the earth, wherein it changes to a light brown Chrysalis, seen in the plate at *(g)*; and at the expiration of three weeks, the Moth appears; this is shewn at *(h)*, where its upper side is described.

The SMALL PEARL BORDER FRITTILLARIA.

THE first appearance of this Fly is about the latter end of May, or soon after the Pearl Borders; its haunts are always in woods, where it skims swiftly over the tops of the grass; they are seen in plenty, yet the Caterpillars have not been hitherto discovered. The upper side is seen at *(i)*, and the under at *(k)*.

The CLIFDEN NONPAREILE.

WE are now come to treat of one of the scarcest Moths of the English production; although indeed much cannot be said respecting its history, that may with certainty be depended on: for although the Aurelians are naturally industrious through their continued hope of meeting with somthing new, yet in all their researches, but two of the Moths have been taken hitherto; the first at Clifden in Buckinghamshire, in July, and was in the possession of Esquire Lockyer of Ealing; the other is now in the cabinet of Mr. Belliard, of Pall-Mall, from which fine Moth, I have coloured those in the plate: but the Caterpillar and Chrysalis have never been seen in England. Those which I have introduced in the work, are copied from the excellent work of Mr. Admiral, the truth of whose drawings, I should do him great injustice to doubt of. But we are still at a loss for the food of the Caterpillar, for altho' that Gentleman has represented the Chrysalis in one place seemingly spun up in the nettles, yet that is no inference of its being the food of this insect; but if we may judge from appearances only, it should seem the Caterpillar feeds high on some tree, as the Caterpillars of the rest of this class do; which are the Crimson Underwing, and Red Underwing, and appears in the Moth about the same time of the year; neither do I only suppose so from the class to which it belongs altogether, but from the length of the middle or belly legs, as may be seen by the extended leg, near the letter *(a)*, in the plate; for I observe all those Caterpillars have such legs, do feed on some tree, and hold so fast to the branches, that 'tis a very difficult matter to disengage them, without some material damage; of this matter the Green Silver-Lines and Scarce Silver-Lines are instances; likewise the December-Moth, the Spider-Caterpillar, and several others.

If these did not feed high, and so likely to suffer by falls, why should nature make such provision for their safety? the December-Moth, is indeed hairy, which would greatly save it from the hurt it might otherways sustain by a fall, but yet it must be observed that this Caterpillar is perhaps one of the slowest movers of any I know of, and should it fall from the tree to the ground, it would be a doubt whether it would not perish ere it attained to its food again; therefore I am naturally directed to this conclusion from the aforesaid considerations; that it feeds on some high tree, or at least in such places where a fall would be certainly detrimental; the Caterpillar is described at *(a)*, the spining at *(b)*, the Chrysalis at *(c)*; the upper side of the Moth is shewn at *(d)*, and the under at *(e)*.

Plate XXXI

Harris's name for this common moth, the Snout-Egger Likeness, was short-lived, and before long it was known as the Snout, Hypena proboscidalis, *so called from the moth's elongated palps. It occurs over most of the the British Isles, wherever nettles abound and hibernates as a caterpillar.*

The Dark Pearl-bordered Fritillary was the name of this bright little woodland butterfly in the nineteenth century and, long before The Aurelian, *John Ray called it the May Fritillary. Still bearing Harris's name, the Small Pearl-bordered Fritillary,* Boloria selene *may be met with locally over most of Britain, but not in Ireland. Its numbers have declined in recent years. The caterpillar, unknown to the early aurelians, has been described delightfully as 'velvety-looking and a smoky pink in colour' with various lines or stripes. It feeds on heath dog-violet,* Viola canina *and, attached to this plant, the dark blunt chrysalis may be found in May but both the larva, which overwinters, and the pupa are extremely difficult to find.*

Here Harris depicts 'one of the scarcest Moths of the English production', the Clifden Nonpareil, Catocala fraxini. *When Benjamin Wilkes published his* English Moths and Butterflies *about the year 1749, only one of these magnificent moths had been recorded in Britain. This was the one caught at Cliveden in Buckinghamshire (he spelt it Cleifden), which is the specimen referred to by Moses Harris as being in the possession of 'Esquire Lockyer'. The other, Harris said, was in the cabinet of Mr Belliard to whom he dedicated his thirty-ninth plate, but he gives no date or place of capture. The moth is an infrequent immigrant to Britain, never in any numbers, and it is just as rare in its appearance now as it was in Harris's day, although on two occasions it managed to establish itself for some years, once in Kent and at another time in Norfolk; but no resident British colonies are now known. Harris rightly inferred that the caterpillar feeds on the leaves of trees, not nettles, for poplar and aspen are the food plants. The reference to the 'Spider-caterpillar' is undoubtedly to that of the Lobster Moth,* Stauropus fagi, *which is of a unique form with long thoracic legs giving a spider-like impression. The caterpillar which Harris copied from another author appears to bear only a superficial resemblance to that of* fraxini.

Ms Harris ad Vivum fecit

*To the Hon.*ble   *Norborne Berkeley*

This Plate is humbly Dedicated by his most   *Obedient Servant Moses Harris*

To Richd Guy Esqr   this Plate is Humbly
Dedicated by his   Obliged Servant,
         Moses Harris.

Plate XXXII

The Meadow Brown, Maniola jurtina *may be found commonly, sometimes abundantly, throughout the British Isles, wherever there are patches of the larval food plants which are grasses of many kinds. The caterpillar, which is green, feeds at night and so it needs to be searched for after dark. This was probably the reason why Harris found it so extraordinarily elusive. Lewin, in 1795, noted that the eggs of this butterfly were scattered here and there over the grass and earth, thus answering Harris's uncertainty, and this is confirmed by Buckler in* Larvae of British Butterflies and Moths *(1886). Merrett first described this species in 1666.*

This little butterfly is one of the first of the non-hibernating species to gladden the English countryside in the warm days of spring. Known to Harris as the Wood Lady, and later as Lady of the Woods in his English Lepidoptera, *it was a name soon dropped for the more apt, but much less picturesque, Orange-Tip,* Anthocharis cardamines. *It haunts lanes and hedges as Harris says; but it is surprising that he did not know the larval food plant for it feeds on several species of* Cruciferae *such as cuckooflower, hedge mustard and garlic mustard, which are quite common and to which the chrysalis, with which he was familiar, is in due course attached. The butterfly, first figured by Moffet, is a familiar sight over most of the British Isles, except the north of Scotland.*

One of the family known as 'Skippers', this butterfly, the Grizzled Skipper, Pyrgus malvae *occurs generally over most of England and Wales, south of Yorkshire. The description of the caterpillar and chrysalis given by Roesel fits that of the early stages of the Grizzled Skipper, except that the food plants in Britain are mainly wild strawberry and bramble. There is no doubt it does fly in woodlands, but certainly prefers more open situations. The insect which Roesel identified with the Grizzled Skipper was probably his native German one, known as the Mallow Skipper,* Carcharodus alceae, *an error which apparently led to much confusion - for several other authors, such as Wilkes and Donovan, were similarly led astray. Lewin, in 1795, called it the Spotted Skipper and made no mistake about the food plant.*

The MEADOW-BROWN.

ALTHOUGH this Fly is, I believe, the most common among us, yet the Caterpillar is very rarely seen; which is very extraordinary, considering 'tis almost eleven months in that state; the first appearance of the Fly is, as near as can be, about the eleventh of June; the hens when impregnated, cast forth their eggs, but I cannot be certain whether they fix them to the blades of grass, or scatter them loose on the ground: the young Caterpillars, produced by the eggs, feed on the meadow grass, during the summer, and conceal themselves all the cold season; during which time of concealment, they do not eat, nor move till the spring; they then come forth and feed till about the tenth of May, when they are about full fed as at *(a)*; they then hang themselves up by the tail, and change to a short thick green Chrysalis, with dark brown markings, as is exactly described at *(b)*; the female Fly shews her upperside at *(c)*, and the under at *(d)*; the male is described at *(e)*, shewing the underside.

WOOD LADY

ABOUT the 6th of May, is the time when this Fly makes its first appearance; and soon after that, they are seen in plenty in meadows; they fly by hedge sides, and frequently traverse round the field, close by the hedge; they very seldom settle, and when they do, 'tis but for a very short time; I am not certain of their food, for the Caterpillars which have been found proved full fed; and had strayed from their food, to seek a place proper for their transformation. But so far of their history may be depended on, that they lay in the Chrysalis state, during the winter, and the Fly is produced the beginning of May; for so did those which were taken in the Caterpillar state at *(f)*. This Chrysalis with its manner of securing itself, is shewn at *(i)*. The upperside of the female is seen at *(g)*, and the male at *(h)*.

The GRIZZLE.

ABOUT the beginning of May, this Fly makes its first appearance; it delights chiefly in woods, or shrubby places, where fern grows, but neither the Caterpillar or the Chrysalis, have yet been discovered in England; Mr. Roesel says, the Caterpillar was found on the mallow, where it lay concealed on a leaf, inclosed in a web, which was spun over the leaf; it appears short and thick, when not in motion; of a redish grey, or pale brown colour, having a dark stripe down the back, and another along each side, of a lighter hue; it lays in the Chrysalis state eleven days; and the Fly when it appears from the Chrysalis, produces blood, like that of the small Tortoise-shell. I have shewn the female flying at *(l)*, and the male at *(m)*, setting, to shew the underside of the wings.

The POPLAR-HAWK.

Plate XXXIII

THE Caterpillars of this Moth are produced from a small round plelucid egg, of the size and colour seen in the plate at *(c)*; these are disposed by the parent, on the twigs of the poplar, and willow, about the middle of May; she does not leave them together in one place, but distributes them in a promiscuous manner, some on one twig, some on another; for were they all laid in one place, the food on that twig would not be sufficient for them; and consequently, they would be in danger of starving, before they could (being so small) find their way to another twig, or branch; they are full fed in September, and appear of the size, and form of that at *(a)*, of a light blue green, having seven diagonal marks on the sides of a pale yellow colour, the last of which ends at the point of the horn at the tail; the breathing holes on the sides are red; the feet, at the head part, are of a rose colour; the rest of the legs are tipt with yellow: the Caterpillar is likewise, covered all over with a Shagreen-like skin; and when at rest, sits in the position as seen at *(b)*; it now goes into the Earth, where it changes into a brown Chrysalis as seen at *(e)*, without any web, or spinning, in which state it lies till about the beginning of May, and then appears in the Moth state, which is described at *(g)*, flying with its wings spread, to shew the upperside; this is a female: the male is described at *(d)*, siting in its natural position; the underside is described at *(f)*.

SMALL MAGPIE LIKENESS.

THIS Caterpillar feeds on the great stinging nettle, spun up in a leaf, as described at *(l)*; it is of a transparent blue-green colour, as seen at *(h)*; about the end of June it changes to a long taper Chrysalis, within the inclosed leaf, which Chrysalis is of a dark-brown, red colour, and of a fine polish all over, as shewn at *(i)*; the Moth is seen flying at *(k)*; and is much of the colour and changability of mother of pearl.

GREY SCALLOP'D BARR.

THE Moth *(m)* was caught near Hallifax in Yorkshire, by Mr. Bolton, who informs me that he took it in the evening, on the moor; and is the first of the kind I have ever seen.

SHADED BROAD BARR.

SENT me by the same gentleman, who informs me he took it in the months of May and June; seen in the plate at *(n)*.

SMALL YELLOW UNDERWING.

THIS Moth is seen in plenty about the latter end of May, flying in long grass by hedge sides, but neither the Caterpillar or Chrysalis have as yet been discovered. I have shewn the Moth at *(o)*.

The Poplar Hawk-moth, Laothoe populi *is found all over the British Isles, and is the commonest Hawk-moth to be seen there. Possessing no tongue, it cannot feed, and is not therefore a visitor to the garden.*

The Small Magpie Likeness is now popularly known as the Mother of Pearl, Pleuroptya ruralis, *having a similar shell-like colour and variation, as Harris observed. It is to be found everywhere in the British Isles where nettles abound.*

It is not surprising that Harris had not met before the Grey Scalloped Bar, Dyscia fagaria *for it is a moorland insect, the caterpillar feeding on heathers and heaths. Mainly a moth of the northern counties, it is found in some localities in southern England, but only rarely.*

The moth sent from Mr Bolton, presumably taken in Yorkshire, may have confused Moses Harris for it is not easy to separate this insect from several similar species. The moth which is now known as the Shaded Broad-Bar, Scotopteryx chenopodiata *has its flight time in July and August, and a close examination of Harris's figure suggests his insect may be the Lead Belle,* S. mucronata scotica, *which flies earlier than the Shaded Broad-Bar.*

In May and June this busy little moth, the Small Yellow Underwing, may be seen visiting wayside and meadow flowers in bright sunshine. The caterpillar of this insect, the scientific name of which is Panemeria tenebrata, *feeds on the seed capsules of the common mouse-ear and can be found over most of Britain in rather scattered localities.*

PL. XXXIII.

To Mr. Emanuel Mendes da Costa,
Librarian and Museum Keeper to the Honble The ROYAL SOCIETY;
This Plate is Humbly Dedicated, by his most Obedient Servt. Moses Harris.

Mr Harris Fecit

Plate XXXIV

Silver Washed FRETILLARIA.

THIS is the largest of all the Fertillarias, and as the rest, so this is always found in or near woods; the time of their appearance in the fly state is about the 10th of June. I have sought for the Caterpillar very carefully, but have never yet been successful. Mr. Rosell, who gave the figures of the Caterpillar and Chrysalis in his work, says it feeds on nettles in private recesses of woods; I know not what may be their food in that country, but have great reason to believe they do not eat it here; and in those woods, where I have always found the flys in great plenty, I have not been able to find any nettles, neither in the woods nor their environs. I did once discover some nettles, in the marshy part of the wood where nothing but small shrubby bushes grew, but I never saw an instance of that kind before; and I am sure 'tis not usual for that plant to grow in woods; I have, however, searched for many years on these nettles, which I could find grew nearest to woods; but all in vain. I have given the figures of the Caterpillar and Chrysalis, from Mr. Rossell's work, the Caterpillar at (k), and the Chrysalis at (l). This insect is in the Caterpillar state during the winter, and changes to a Chrysalis in May, and lays in that state near three weeks, as all the rest of the Fretillaries do, this we are easily informed of from the time of year in which it appears in the fly state, altho' we cannot with any certainty tell the food; though, indeed, I have great reason to suspect the blamble, as they are so remarkably fond of setting on its blossoms. The female is seen flying at (m), shewing the upper side, and the male at (n), discovering the under: the difference between the two is easily seen; the male being redder on the upper side, and three of the middle fibres of the upper wing lay very high, are black, and as it were hairy; for the better understanding of which, compare the two wings together, that of the male with the female.

GREEN FORESTER

COMMON meadow sorrel is the food of this Caterpillar, on which it may be found feeding; the latter end of April nearly full fed, they change into a Chrysalis, of a light brown colour, within a pretty strong web, which appears double, or rather one within another the beginning of May, and the Moth appears the latter end of that month; I have shewn the Caterpillar, as full fed at (a), which on being touched, falls down and turns itself up in a curled form, as at (b); the spinning is described at (c), and the Chrysalis at (d). The hen Moth I have shewn flying at (e), whose upper wings are of a yellowish green, and appear something like sattin; the cock is less, and of a blueish green; the under side of which Moth is described at (f).

L. MOTH.

PROCEEDS from a Luper Caterpillar, somewhat hairy, and of a brownish green colour; it feeds on the goosberry-bush, and sometimes on the currant-bush, but the former appears to be its favourite food; it is all winter in the Caterpillar state, and becomes full fed, as at (g); the latter end of May changes in the earth to a brown Chrysalis, which is described at (h), and the Moth appears the beginning of June, which is described at (i), shewing the upper side of its wings.

COPPER.

THE small Butterfly at (p), is one of those whose history is not compleat; but I shall say so much of it, as I can prove, from circumstances, to be true relating to the Fly. I have several times taken this Fly the beginning of April in a very shattered condition; from which I infer, that it remains in that state during the winter. About the latter end of June another brood appears fresh and new, which, no doubt, proceed from those which appear in April; from the June brood another proceeds, which appear in the Fly about the latter end of August, go thro' the winter in that state, and are those which are seen in April following; they fly in woods and meadows, settling by path sides, on the tops of grass.

DINGEY SKIPPER

THIS Butterfly appears fresh the sixth of May; they fly very low over the tops of the grass; I don't perceive them to be fond of any place in particular; they fly in great plenty in woods, meadows, heaths, &c. &c. and are described in the plate at (o).

Mottled ORANGE

Plate XXXV

THE food of this Caterpillar, is the pith within the stalk of the burdock, wherein it may be found full fed in the middle of July, by tearing or spliting the stalk from top to bottom. The manner in which the Caterpillar secures itself is very artfully contrived; for the stalk of the dock, by the inside being so much destroy'd, is rendered incapable of defending it from the researches of the Ichneumon, being so fractured and perforated; for I have often watched, and seen with what diligence those animals seek the Caterpillars; they will examine the stalk from top to bottom, searching in every hole in less than five minutes; but the Caterpillar, as if wary of so dangerous an enemy, generally keeps every avenue stopped with its own dung; yet, nevertheless, in despite of his care they do often get at them, and destroy them, as I have often found the Chrysalides of the Ichneumon laying in the stalk of the dock, close by the expiring Caterpillar; but when the Caterpillar has become full fed, as at (a), he then begins to prepare a place of security, sufficient to baffle the art of his enemies; which is thus performed, he retires downward to the bottom part of the stalk, where he is sure to meet with a great quantity of his dung, which he gets through, or rather under, till he almost comes to the bottom; here he begins to shape his cell, making it capacious enough to contain him with ease, spinning the dung together with a slight web, that it may not fall in upon him; his next business is to gnaw a hole through the stalk to the outside, large enough to admit his escape, when in the Moth state: this hole the Caterpillar covers over with a thin, but strong web to preserve him from the attempts of the Ichneumon, as seen at (c), who otherwise might possibly get in, when he is in that tender state of changing into the Chrysalis; within this cell he changes to the Chrysalis, shewn at (b); and the Moth appears the latter end of August; the hen is shewn at (d), and the cock at (e).

HALF MOURNER

THE Caterpillar feeds on oak leaves, on which it may be found by beating at the latter end of May, nearly full fed, as at (f); it changes to a short, thick Chrysalis, the beginning of June, in a spinning against the leaf, and the Moth appears the latter end, or about the 20th of August, which is shewn at (g).

RINGLET

THIS Butterfly generally appears about the 20th June, flying in company with the Meadow Brown, so that 'tis very difficult for a young Aurelian to know one from the other; for besides being nearly of a colour, their manner of flying is so exactly alike, that an experienced Aurelian is often mistaken; the best mark to know them is by their darkness of colour, which causes them to appear almost black as they fly; the Caterpillar or Chrysalis I have never yet seen, but make not the least doubt of its being much the same with the Meadow Brown, and are all the winter in the Caterpillar state: I have described the fly at (h), shewing the under side; this is the female; the male is much like it, but less; the upper side I thought not worth the representing, being all over of a dark brown colour, without the least variety of marking, except one small dark spot in each wing.

CLOUDED YELLOW

THIS pretty Moth is taken by beating the whitehorn hedges about the middle of June; for which reason, the Caterpillar is supposed to feed thereon: I do remember breeding this Moth in a cage, wherein I had put a number of small green Caterpillers, beat in May, and some of them Lupers; but to say, which was the Caterpillar that produced the Moth I cannot, but presume it was one of the Lupers by the form of the Fly, which is seen at (l), shewing the upper side of his wings.

SEVEN SPOT ERMINE

I Received this Moth with many others, from a friend in Yorkshire, who informs me he took it in May, and that it is there very scarce, but in these more southern parts has never yet been discovered by anybody; therefore is esteemed as a great curiosity.

ORANGE UNDERWING

IT appears in the Moth state about the latter end of March; but the best time to go in pursuit of them is the first of April; about which time they are very fresh and in great plenty; that is to say, at the place where they are taken, which is on the S.E. side of Hornsey-Wood, facing the river; the best time to go is when the wind is at W. or N. W. which blows them out of the wood a great way into the meadow, and frequently very high in the air; so that in your pursuit, you must observe to keep under them, waiting for their dropping down, which they certainly will to get under the wind, in order for obtaining the wood; but should you not be under them when they descend, and catch them in your net directly, they will certainly treat you with a smart chase; 'tis presumed their Caterpillars feeds on the Arbeel, which grows plentifully in that wood, that they are in the Caterpillar state during the summer, and lay in the Chrysalis state all winter. I have shewn the Moth at (i).

The moth called Mottled Orange is now known as the Frosted Orange, Gortyna flavago. Harris's account of the caterpillar is a wonderful example of his patience in investigation; in addition to burdock it attacks the pith of many plants, such as thistle and foxglove.

This moth, now known as the Oak Lutestring, Cymatophorima diluta was called Half Mourner by Harris and the Poplar Lutestring Likeness by the famous naturalist William Jones of Chelsea. 'Lutestring' had already been incorporated in the name of a fairly similar species, the Poplar Lutestring.

Although unknown to Harris when The Aurelian was written, it was not long before he discovered the food-plants of the Ringlet, Aphantopus hyperantus, as they were correctly stated as 'grasses' in The English Lepidoptera a few years later.

The leaves of any sort of wild or cultivated rose provide food for this bright little moth, the Clouded Yellow now called the Barred Yellow, Cidaria fulvata, to be found in most places in the British Isles.

Harris's name for this moth was not inappropriate, but he was unaware that the male of this species, which he called the Seven Spot Ermine, has a quite different appearance; this would have puzzled him greatly. It is a dark brown or blackish colour, and smaller than the female he portrays. The Muslin Moth, its scientific name Diaphora mendica has superseded Harris's name, which was misleading because the number of black spots on the wings of the female varies.

There are two moths, the Orange Underwing, Archiearis parthenias and A. notha, the Light Orange Underwing, which have great similarity. The ones Harris chased so energetically appear to be the latter, for aspen woods are the haunt of this early spring species. A. parthenias caterpillars feed on birch, whereas those of notha are aspen feeders. The Orange Underwing is now a very local insect found mainly in the south of England. Moses Harris's friend Dru Drury had a different method of capturing the 'Orange Underwing' as described in his journal of 1764: 'March 29 — Took many Orange Underwings in Hanging Wood a Confirmation that they are found at more places than Hornsey, NB the best way to take this fly is on a Day windy but not boisterous chuse a part of the wood cleared of the undergrowth — let one beat with one stroke of a club the body of a young arbeal or alder left singly by the woodmen whilst another stands with a net under the wind to take them the moment they fly.'

To the Rev.d Mr Will.m Ray
This Plate is humbly Dedicated by his most humble Obliged Serv.t
MOSES HARRIS

BONO VINCE MALUS

Plate XXXVI

SWALLOW-TAIL

The plate of the Swallowtail Butterfly, *Papilio machaon* is dedicated to the Reverend Mr William Ray, who sent him the larvae from near Bristol, and one assumes they were collected in that area. For although this most exciting butterfly is now restricted to certain parts of fenland Norfolk, it formerly had a much wider range but nowhere, except in the Fens, has it ever been common. It is now in danger of extinction and is protected by law. *Moses Harris, in his* English Lepidoptera *gave 'meadows, Bristol and Westram' as the 'Haunts in the winged state' of this rare 'fly', but he made no reference to the type of country to which it is now confined; neither did Lewin in 1795.*

Old Thomas Moffet first figured it, and Petiver in 1717 said it was 'caught about London and divers counties in England, but rarely'. He called it the 'Royal William', a name used by some aurelians and which, said Harris in 1776, 'was probably a compliment to His Royal Highness, William Duke of Cumberland, who was popular for his defeat of the rebels in 1745'.

ON the 10th of September, 1760, I received from the Rev. Mr. Ray, of Redland, near Bristol, a box containing twelve Swallow-tailed Caterpillars, three of which were dead, I supposed by the shaking of the box in the carriage, and their close confinement; the remainder were in good health, and each in its last skin, except one Caterpillar which was less than the rest, and is seen in the plate at *(a)*. I fed them on carrot-greens, which they eat freely; in about three days some became full fed, and appeared then as represented at *(b)*; they then forsook the food, getting on the side and corners of the cage, where they remained fixed and motionless; after resting two days in this manner, I found one of them had fastened his tail to the wood by a small white spinning, and was very busy in making a fine white silken thread, the ends of which were fixed, one on each side, pretty near the head. View the figure at *(c)*. I could not help admiring with what care and pains he worked to make it strong, rubbing with his mouth backward and forward, with such a motion as the shoemakers use in waxing their ends. When finished, he put his head under, or through it, and the thread then fell a-cross his back; but the thread now appear'd too big for him; after this he remained two days more, during which time he shortened and grew thicker, and at length changed into the Chrysalis, which is represented at *(d)*, where it may be observed that the silken string, which appeared too big before, is now filled up by the thickness of the Chrysalis. I have shewn another figure of the Chrysalis at *(e)*; for the better describing the front part, whereby the flatness of the tail may be seen. The Flies appeared the middle of May. The upper-side the female is shewn at *(f)*, and the under-side the male at *(g)*; from this brood which appears in May, proceeds another which appears in the Fly state in August; and the brood proceeding from these goes through the winter in the Chrysalis state, and appears in the Fly in May.

BEE-TIGER

Plate XXXVII

BY some called Caput Mortuum, or Dead-Head, from the mark on the back, which much resembles a dead Scul; others call it the Jasmine Hawk; and by Mr. Ray, of Redland, the Pottatoe Hawk; but the Aurelian society chuse to distinguish it by the name of the Bee-Tiger, but for what reason I know not; perhaps there is not another insect in the world so general as this, being produced in almost every country and climate, their food differing according to the place. I have seen those which have come from the East and West-Indies, others from Africa, Italy, Germany, and without any visable or material difference between them, only, I think, our European ones the largest; but as I am treating of the English insects only, I shall proceed to give you an account how they breed with us. The Caterpillar which I have represented in this plate, sometimes feeds on jasamine, and often on elder. But Mr. Ray is of an opinion their principal food, at least in this island, is the pottatoe plant, that gentleman having found them in plenty several succeeding years feeding on that plant. I shall therefore proceed to give you his account of them, as near as I can, in his own words, which I think so much to the purpose, that to say more on this insect is almost needless, at least in this treatise, where I am obliged to be as concise as the nature of the thing will admit of.

'After hunting some years, all the jasamine and elder, (which I learnt the Caterpillar of this Moth fed on,) that I could hear of, without success, I dispaired of ever procuring it; but being informed by some gardeners of Somerton, in the county of Somerset, that their potatoes were infested by a Large-Grub, as they called it, I desired them to send me some of them; and accordingly, in a few days they remitted me four, which turn'd out to be the very thing I had been so long in quest of.' 'Upon this, I dispatched my servant immediately, to search the fields where these were found, and he brought me three more: This was the first week in September, old stile; and the frost setting in that year pretty early, put an end to my further pursuit after them for that season; but the next, about the same time, I repeated it, and was equally successful; as I was likewise the third year: And I have been informed, they have been taken there since I left that country. They have likewise been found in the neighbourhood of Bristol, feeding on the same plant. All these Caterpillars were pretty large when they came to my hand, and remained with me but a few days, when they went into the ground, where they form'd tombs, like those of the Private, Eyed-Hawk, &c. and about the middle, or latter end of June following, came into the Moth, which has this circumstance particular to it; that when you approach or disturb it, it chirps like a bird. The Caterpillar is a large feeder, and requires to be frequently supplied with fresh food: The place where they were found, is on that side Somerton, next Kingsedgemore: and the soil is a red loom.'

I observe Mr. Wilks says, the Caterpillar goes into the ground in July, and the Moth is bred in October; but all that I know concerning the history of this Moth, and accounts received from others, exactly agree with Mr. Ray's; and particularly a gentleman, who bred a Moth, which I had afterwards of Mr. Peter Colinson, expressly says, in his letter to that gentleman, that the Caterpillar buried itself in the earth, (prepared in a pot for it) the 4th of September, and the Moth appeared on the 5th of July following. As to the method the hen Moth takes in depositing her eggs, 'tis not at all to be doubted, but she lays them in the same manner as the rest of her Class, *viz*. Eyed Hawk, Privet, &c. &c.

The resemblance to the skull of the marking on the thorax gave rise to the scientific name of Acherontia atropos *for the Death's-head Hawk-moth, from the underworld river Acheron and the Fate who cuts the thread of life.*
To Harris it was the Bee-tiger, and it has now been established beyond doubt that it enters bee-hives in order to steal honey and, to a large extent, it seems mysteriously to avoid being stung. This propensity was apparently unknown to Harris, hence the questioning of its name. However, a few years later he referred to it by the common name by which it is still known. To add to the curiosity of this magnificent insect, it has the unique power of producing an audible squeak, noticed by Harris's correspondent. Powerful in flight, it migrates to Britain from the Continent, but the climate does not allow it to complete its life cycle there in the wild, and although larvae and pupae are sometimes found, they must be reared artificially if moths are to be produced.

To my Ingenous Friend and Benefactor Mr Dru Drury.
This Plate is most Humbly Dedicated by his Obliged Servant Moses Harris

Mo.ˢ Harris ad Viv

At the Expence of D.ᵣ Fothergill, to whom
this Plate is Inscribed by his Obliged Servant Mofes Harris.

Plate XXXVIII

This plate showing Cerura vinula *is dedicated to Dr Fothergill but, unlike the other dedications, no coat of arms is displayed, for Fothergill was a Quaker whose custom it was not to display such an outward sign of distinction, or to accept honours. The caterpillar of the Puss Moth is probably the best known of any of the native British species because of its curious appearance and the ability to eject Harris's 'thin liquor', which is in fact formic acid capable of being expelled up to a distance of about two feet. The Linnean name* vinula *is hardly appropriate if, as so often asserted, the moth was named for that reason! However, the erudite author, who loved and named the caterpillar, was so delighted with its rich wine colouring that it was emphasized in his description, published in 1634. The Puss moth, so called from earliest times, is named, it is said, for its downy appearance but the caterpillar, so extraordinary, hasn't earned a separate sobriquet, which it so thoroughly deserves! It is found over most of the British Isles.*

The name Large Ermine was changed during the nineteenth century to White Ermine, Spilosoma lubricipeda *and is very similar, except for its colour, to the Buff Ermine in the same genus (shown on Plate XVII). It is a common moth over most of the British Isles.*

Harris's candid ignorance of the early stages of the Heath Fritillary, Mellicta athalia, *which he calls the Pearl Border Likeness, is excusable for then, as now, it was probably local — although he does mention that in certain places it could be taken in plenty. However, it has become increasingly scarce and is now found in a few places in southern England only — it is so much in danger of extinction in Britain that, as with the last butterfly depicted by Harris, the Swallowtail on Plate XXXVI, it is under legal protection. The account given by Harris of this butterfly is mostly correct, but he was wrongly advised regarding the larval food plant, for this is usually cow-wheat, and the chrysalis is dark-brownish with a rather speckled appearance. Petiver in 1717 called this butterfly the Straw May Fritillary. The letter references in the plate and text are not in accord, which indicates a revision in one or the other of these.*

PUSS

THE Caterpillar feeds on aspin, and several sorts of willow, on which the parent Moth lays her eggs about the middle of June, disposing them in different places, or parts of the tree: they are of a redish-brown colour, and appear a little flatted: the young Caterpillars, when they appear from the eggs, are of a very dark-blue, or black colour, and their tails remarkably long, as at *(a)*: But as they shift their skins, and grow larger, they appear lighter, and their tails shorter in proportion. When they have arrived to a tolerable size, their form being then more conspicuous; they appear of an extraordinary shape; the front or head part being square in the center of which the head lies almost concealed: On the two corners, over the head, are two round black spots, one on each corner, which at first sight appear like eyes; on the back it hath a rising or angle, which is on the third joint from the head, the top or point of which, is red. Along each side, from the head to the tail part, there runs a broadish white line, which joins or meets together on the point of the rising of the back, and then divide again, descending with a sloop down each side, till they arrive at the tail part. This Caterpillar hath no holders behind, nor does it ever let its hind part touch the ground in creeping; but instead thereof, there are two tails, or tubes, thro' which, when provoked, he thrusts out two red arrows of a tender elastick substance, as at *(b)*; and upon a repeated insult, ejects from thence a thin liquor, which they have often squirted in the face of them which attempt to injure them. The Caterpillar is a fine green beneath, and of a light blue-green above the white lines, and on the edge of the white line toward the middle, shaded as it were, with a redish-purple: But the figures in the plate will best describe it. It is somewhat remarkable, that when the Caterpillar is near full fed, he looses the power of putting forth these darts, by the drying up, or shriveling of the tubes; See the Caterpillar, as at *(c)*. When full fed, as at *(c)*, they spin a hard case against the bark of the tree, &c. wherein they change to the Chrysalis, which is seen at *(d)*; this happens about the latter end of August, and the Moth appears about the latter end of May: The Hen is shown flying at *(e)*, and the Cock setting at *(f)*.

LARGE ERMINE

ELDER seems to be the most beloved food of the Caterpillar, as I have found the eggs thereon, but never on anything else, although the Caterpillar will almost eat any thing. The eggs are round, and of a shining light-green colour; these are laid upon the leaf in regular form, like those of the Tiger, in May; and the young Caterpillars, when they appear from the shells, feed on the leaf; but as their chaps are too tender to bite thro' the whole leaf, they only at first eat the fleshy part, leaving the membranous skin intire, as at *(g)*. The Caterpillar, when full fed, appears as at *(a)*, which is hairy, with an orange coloured stripe down the back, or pulse; they change to a Chrysalis, within a web, toward the latter end of July, and the Moth appears the middle of May following: the Chrysalis is shewn at *(e)*; the Hen Moth is described at *(d)*, in a sitting position; and the male is seen flying at *(e)*.

PEARL BORDER LIKENESS

NEITHER the Caterpillar nor Chrysalis of this Fly has as yet been discovered: some indeed affirm, that it feeds on heath; that the Caterpillar is black, and full of springey bristles; the Chrysalis is black, short, and thick, and changes to Chrysalis, hanging by the tail: And indeed, there appears some probability in this account, as it extremely well answers to the nature of this class. They appear about the middle of June, and may be taken in plenty flying on heaths, near woods, and in woods where heath grows: I have shewn the underside at *(k)*, and upper at *(b)*, which is the male.

Buff TIP'D

Plate XXXIX

THIS Caterpillar feeds on oak and most kind of willow, and is taken by beating the boughs of oak, &c. the latter end of August, or the beginning of September: It is, when full fed, near three inches long, of a dark olive colour and divided into squares by transverse stripes of yellow, which gives it something of the appearance of a Scotch plaid; they bury themselves in the Earth in the month of September, and change into a pretty short, thick Chrysalis, of a dirty brown colour, and the Moth appears the latter end of May. The Caterpillar is seen at *(a)*, the Chrysalis at *(b)*; the male Moth is described in its sitting posture, at *(c)*, and the hen as flying at *(d)*. I once found some eggs of the Buff Tip'd Moth, which were of a greenish colour, quite round, and were laid on a leaf in perfect and regular order, by the side of each other, as appears in the plate at *(e)*; after I had kept them some time, as a day or two, to my surprize, a small Ichneumon appeared from each egg, of the exact shape and colour, as described at *(f)*, which is magnified on purpose, the better to shew its formation. This is the first instance I ever saw of the eggs of any Moth, or fly, being perforated, by what we distinguish by the name Ichmeumon.

LARGE YELLOW UNDERWING

THE Caterpillar feeds in gardens on many different plants, and in the fields, on grass, at the roots of which, it buries itself in the earth during the winter; and in the month of April, it appears again on the surface, feeding on grass, till it becomes full fed, as at *(g)*, which happens about the middle of May, when it goes into the earth; and after making itself a kind of tomb, changes to a light brown Chrysalis, as seen at *(h)*; and the Moth appears in June. I have described the Moth with the two wings spread at *(i)*, to shew the under wings; and at *(k)*, to the position in which it sets when at rest.

The leaves of almost any kind of tree or shrub may be eaten by the caterpillar of the Buff-tip, Phalera bucephala *which, until it is nearly full-grown, feeds in companies. Its specific name* bucephala *was given to it because of the 'bull-headed' appearance of the moth when resting, but more worthy of note is the remarkable resemblance to a piece of broken stick when in that position. Harris's account of the attack on the eggs of this moth by one of the 'insectophagi' (as he calls them in Plate I) must be one of the earliest recorded. The Buff-tip is common in the London area, as well as over most of the British Isles.*

Sometimes this moth, Noctua pronuba, *with its bright yellow underwings and rapid flight, may be disturbed in huge numbers from meadows in the summertime. It is presumed these are immigrants but the Large Yellow Underwing is always a very common insect. The caterpillar, which can be a pest in the garden, has the habit of burying itself just below ground surface during the day, and feeding on a wide variety of plants at night, and will continue to eat during the winter when conditions are suitable.*

Ms Harris ad Vivum May 1 1766

To Charles Belliard Esq.ʳ this Plate is most Humbly Dedicated
by his Obliged Servant Moses Harris.

PL. XL

To Peter Collinson Esq.ʳ F.R.S. This Plate is Humbly
Dedicated by his most Obliged & Obedient Servant MOSES HARRIS.

Plate XL

The Scarlet Tiger, Callimorpha dominula is a very local moth which may be found, although rarely, in England and Wales, but only in their southern and western parts. It also occurs infrequently in two places in Kent. It is interesting to note that it was in this county, over two centuries ago, that the brilliant insect was discovered by Moses Harris. It seems to prefer rather damp and marshy places but it flies also in bright sunshine. Harris's observation that the eggs are laid on the food plant is significant, for this has been confirmed by the most recent authorities in The Moths and Butterflies of Great Britain and Ireland *(Volume 9, 1979). This is at variance with the previously accepted belief that they were scattered at random, having no adhesive to attach them to a plant, as stated in Stokoe and Stovin's* Caterpillars of British Moths, including the Eggs, Chrysalids and Food-plants *(1958).*

Many years ago the Pearl-bordered Fritillary, Boloria euphrosyne was known to Ray as the 'April Fritillary', a name still used by Lewin in 1795, although (by the modern calendar) May and June are the months when it can be seen flying in its woodland haunts. Lewin reported having seen it on the wing as early as April 12, but this must have been very exceptional. The food plant is heath dog-violet, which is the same as for the Small Pearl-bordered Fritillary (which is shown on Plate XXXI) and both overwinter as larvae. The caterpillar is black when fully grown, with various stripes, and has also a velvety appearance. The range of this butterfly is widespread in Great Britain, becoming scarce further north, and it occurs in one Irish locality.

Scarlet TIGER

THIS Caterpillar is to be found on houndstongue, nettles, hoarhound, &c. but houndstongue is supposed to be the proper food. The best place to obtain this Caterpillar, is, at Charlton in Kent, down in the Chalk Dell, near the half-way house to Woolwich; and on beating the nettles which grow on the sides of banks, or other eminences, they will roll down in plenty. The best time to search for them, is the latter end of April, but it will not be too late the beginning of May: They are not to be taken any where else, that I know of, however, not in such plenty. I did once pick up one at Erith, in Kent, and I endeavoured to discover some more, but searched in vain for some time. They are black, with a double spotted stripe down the back, composed of white and yellow spots, the form of which spots, is better described in the plate: It is spotted likewise with the same coloured spots on the sides. The head is black, and the body covered with a light yellowish hair, of a middling length: When full fed, as at *(a)*, which happens about the middle of May, it spins itself up in a slight web, of a lightish colour, and changes to the Chrysalis, which is of a deep red colour, near to black, as seen at *(b)*; and the Moth appears in thirty days. The Moth is shewn flying at *(c)*, the better to discover the under wings; and at *(d)*, as sitting in its natural position; the eggs are of a deep gold colour, and adhere to the place whereon they are laid; and the Caterpillars, when hatch'd from the eggs, continue in that state during the winter.

PEARL BORDER FRITTILLARIA

THIS Fly is taken in woods about the 11th of May, as I never remember to have seen any before that time; they fly low, and are generally taken in tollerable plenty, but the Caterpillar and Chrysalis, I have never yet been so lucky as to discover. I have shewn it at *(e)*, displaying the upper side; and at *(f)*, sitting in a position which shews the under side.

Angle SHADES

Plate XLI

THIS Caterpillar feeds on nettles, chick-weed, &c. on which it may be found full fed about the 20th April; when it appears, as at *(c)*, large and bulkey, of a fine transparent green, with a darkish mark down the back. They change to Chrysalis within a spinning on the surface of the Earth, and the Moth appears in thirty days. I have described the Chrysalis at *(d)*, of a fine deep glossy red colour, and is remarkable for having two sharp points at its tail: The Moth is seen at *(e)*, shewing its upper side: The eggs proceeding from this brood, which is called the first, produce Caterpillars, which become full fed about the beginning of July, change to their Chrysalides, and appear in the Moth state the middle of September, which is called the latter brood; and the Caterpillars, which proceed from the Eggs of this last brood, remain in that state during the winter, and appear full fed the latter end of April, as above, and are those which produce the first brood.

BROWN HOOK TIP'D

THE Caterpillar of this Moth, is taken by beating the oak, which is, I believe, its only food: About the latter end of September it changes to Chrysalis, in a spinning among the leaves, the end of September, or beginning of October, and the Moth appears in the middle of May: The Caterpillar is seen at *(a)*, in the describing of which I have been the more particular, as its form is so remarkable; the hinder part of the back and tail is a very clear white; the side, of a fine olive green; the belly of a redish brown; as is the protuberance, and the point at the tail; that part of the back behind the head is of a dirty white, or very light brown. The Moth is seen flying at *(b)*, shewing the upper-side of her wings. I have not figured the underside, as it hath no particular markings worth notice.

MASK

THIS is called the Mask-Moth, because it has something of the appearance of a face on the wings: all which I can at present inform you of it is, that it is to be taken about the latter end of May, flying in the grass: Neither the caterpillar nor Chrysalis has as yet been discovered.

SPECKLED WOOD.

THE Caterpillars feed on grass, in the bushy parts of woods, but are very difficult to be seen on account of their colour, which is green, with a narrow light mark along the side, which is dark on the under-side; it hath two points at the tail. It changes into a short thick green Chrysalis hanging by the tail. And the Fly is bred in fourteen days; there are three broods a year, the first in April; the second in June, and the third in August; the Caterpillars produced by the eggs of the August brood continue all winter in that state, and change into Chrysalis the letter end of March; and the flies appear the middle of April: The Caterpillar is seen at *(f)*; the Chrysalis at *(g)*; the upper-side of the female at *(h)*, and its under-side at *(i)*. I have given a figure of the male at *(k)* shewing its under-side.

BARR'D HOOK TIPT.

THE Caterpillar feeds on oak leaves spun, or rather rolled up in the leaf, and is to be found full fed about the latter end of May; it is of a very dark green olive, and better than an inch long; it changes into Chrysalis within the leaf, and the Moth is produced the beginning of July; the Caterpillar is seen at *(l)*; the hen Moth at *(m)*; and the cock at *(n)*, where it is shewn sitting in its natural position.

BROWN CHINA MARK.

THE Caterpillar feeds spun up in the leaves of elder; it is of a fine green colour, prettily marked down the back, as at *(o)*; it changes to a light brown Chrysalis, seen at *(p)*, within the leaf, about the letter end of May, and the Moth appears the letter end of June, which is described at *(q)*: There is another species of this Moth, which I have seen flying in swarms, about rushes in watry and marshy places, towards the middle of June; they appear larger than this, but much paler in colour. I am inclined to think these Caterpillars feed on the bull-rush.

MAID OF HONOUR.

THIS Moth is taken by beating of white-thorn hedges, about the beginning of June, on which it is supposed its Caterpillar feeds; it proceeds from a Luper Caterpillar of a greenish colour, and changes to Chrysalis within a spinning; the Chrysalis is of a brown colour, very short and thick. I had not the opportunity of giving the figures of the Caterpillar and Chrysalis in the Work, having, by some means or other, mislaid or lost the drawings. The Moth is shewn flying at *(r)*.

Had Harris portrayed this moth, Phlogophora meticulosa *or the* Angle Shades *in its attitude of rest, he would have shown an excellent example of insect camouflage. The upper wings are drawn about the body, and wrinkled throughout their length. This, combined with its obliterative shades of brown, give a remarkable resemblance to that of a crumpled leaf.*

Harris was right in believing that oak was the sole food plant of this little moth, called the Oak Hook-tip, Drepana binaria. *However, he seemed unaware this little insect is double-brooded.*

In The Aurelian *Harris calls this moth the Mask. In the* English Lepidoptera *(1775) he refers to it as the Shipton, now amended to the popular name which has been in use for a long time, Mother Shipton,* Callistege mi. *The face-like markings on the wings were supposed to resemble a witch of that name who achieved notoriety in Tudor days.*

It is unusual for any British butterfly to overwinter either as a pupa or as a larva, yet this happens with the Speckled Wood, Pararge aegeria, *still known by the name Harris called it.*

The Barr'd Hook Tipt is not to be confused with the moth we know as the Barred Hook-tip, of the same genus as the Oak Hook-tip on this plate. Some earlier authors' identify Harris's moth as Lozotaenia obliquana. *They are clearly mistaken, for the most likely insect which Harris illustrated is known as* Choristoneura hebenstreitella, *a large Tortrix moth found mainly in woods in the south of England. However, it is not possible to identify the moth with certainty; another which is common throughout most of the British Isles is* Pandemis cerasana, *which it also resembles.*

The prettily marked Pyralid moth, known to Harris as the Brown China Mark, has no popular name today but is identified as Eurrhypara coronata. *The similar but larger species which he noticed was probably the Beautiful China Mark,* Parapoynx stagnata, *which is common in the London area but prefers marshy watery places.*

Harris admits to carelessness in losing his drawings of the caterpillar of the Maid of Honour. This is a pity because it has a curious habit of covering itself with tiny pieces of food-stuff as soon as it hatches and, when feeding again in early spring, camouflages itself with the scales of the oak buds, its food plant. We now call this moth by the graceless name, the Blotched Emerald, Comibaena bajularia.

Mo.ˢ Harris Feeit Oct.ʳ 21 1763

To Mͬ Andrew Peter Dupont,

This Plate is Humbly Dedicated by his most Obedient Serv.ᵗ Mofes Harris.

PL. XLII

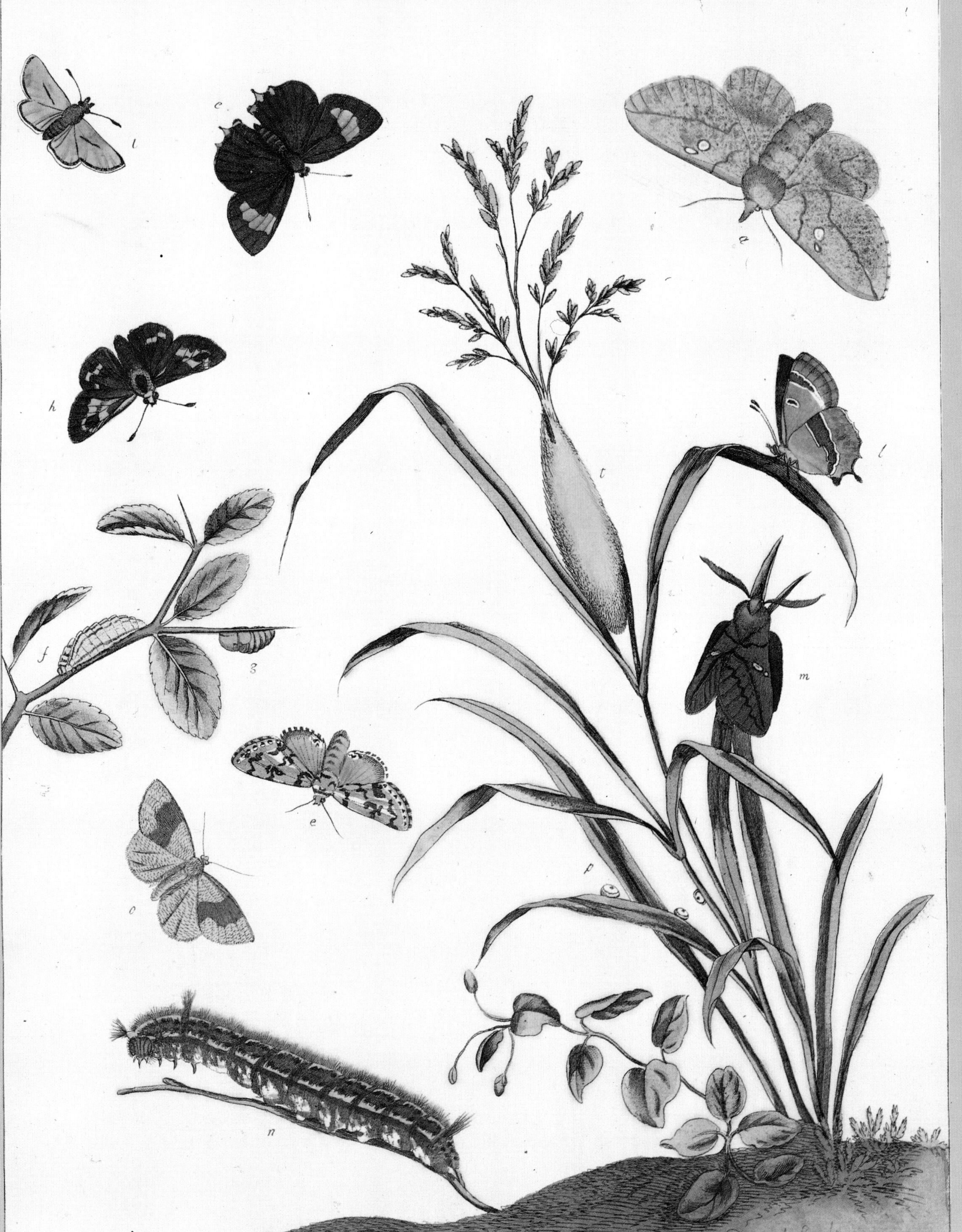

Plate XLII

The caterpillar of this large common moth sips from drops of water giving rise to its popular name. In 1662 Jan Goedart remarked on its thirst, commenting that 'The Drinker' was the name by which it was known then, as now, (reflected in its scientific name Philudoria potatoria*).*

In 1702 John Ray took the first recorded specimen of the Brown Hairstreak, Thecla betulae, *the largest of the British Hairstreaks, so named from the markings on the underwings. It is found mainly in the southern half of England in woodlands and hedgerows, but never abundantly. Either Harris or the printer made an error over the food of the caterpillar; it should read 'blackthorn' not 'buckthorn'.*

Harris's 'Large Skipper' is now called the Silver-spotted Skipper, Hesperia comma, *and he correctly attributed to it the Linnean specific name,* comma. *In* The English Lepidoptera *(1775), he gave it the name Pearl Skipper but this was not generally adopted. The Silver-spotted Skipper was recorded in the last century near London, and Harris said it could be taken on swampy ground near Ealing in Middlesex. Now quite scarce, it is confined to chalk hills and downs of southern England. The dull olive-green caterpillar feeds on sheep's-fescue grass,* Festuca ovina.

Perhaps Harris was influenced by Petiver, who called the male Small Skipper the 'Spotless Hag' (Hog) and thought the female a different species. He believed the female had the 'black stroke' on the upper wings, whereas the male carries this distinguishing mark. To early aurelians 'Skippers' were 'Hogs'. The light green caterpillar of the Small Skipper, Thymelicus sylvestris *feeds on grasses, and the butterfly is common over most of southern and midland England and Wales.*

The Scarce Merveille du Jour, Moma alpium *is found only in scattered localities in southern England. Black markings on beautiful green wings adorn this very handsome moth. The attractive caterpillar is black with three yellowish blotches on the back and a pair of bright red warts on each segment. With no drawing before him, Harris's memory was much at fault when he said the caterpillar was green.*

The Freckled Broad Bar, now called the Barred Umber, Plagodis pulveraria *may be found in woodlands over most of the British Isles, but is never very common. Harris found the caterpillar on oak but birch and sallow are its usual food plants. It is coloured reddish-brown and mottled with yellow.*

The DRINKER.

THE *eggs* seen at *(p)*, are deposited by the female on a large coarse kind of grass which commonly grows under hedges, to the stalks and leaves thereof the eggs adhere, being fixed by a kind of gum not disolvable by water; they are about the bigness of a hemp seed, of a whitish colour, having a spot on the middle of the upper-side, which is incircled with a black ring; they are hatched about the beginning of July, and the young Caterpillars remain in that state during the winter; in spring they come forth from their secret places, and feed till the letter end of May, when they become full fed and appear of the size and form as at *(n)*: They then spin themselves up in long silken cases, or bags, of a buff colour, wherein they change to brown Chrysalides, which are round or blunt at each end, and the Moth appears at the expiration of one month. The male, seen at *(m)*, differs from the female both in size and colour, as he is of a dark red brown colour, having the antenna very long and broad, but the female is of a buff colour, and the antenna very thin and narrow, as seen at *(a)*. They fly in the evening.

The BROWN HAIR STREAK.

BUCKTHORN is the food of the Caterpillar, which produces this pretty Fly: The Caterpillar, which is figured at *(f)*, is green, and of the same form as that of the Purple Hair Streak, but much larger; it is full fed about the beginning of July, when it fixeth itself to a twig by the tail, and having a kind of brace round the middle, changeth to the Chrysalis as seen at *(g)*, and the papilio appears the beginning of August. The female is remarkable for having two large orange coloured spots, one on each of the superior wings on the upper-side: The male is on this side intirely brown. The under-side is seen at *(l)*. They fly in lanes, delighting to play about and settle on the tops of hedges.

LARGE SKIPPER.

THE Caterpillar of this Fly hath never yet been discovered in this land. They delight to fly in woods, and lanes near woods; their actions are somewhat remarkable, and not unworthy notice, for when ever they settle, which is very frequent, as they are never long on the wing, they are sure to turn half-way round, so that if they settle with their heads from us, they turn till their heads are toward us, and sometimes till they are quite round. When on the wing, they have a kind of skipping motion, which is affected by reason of their closing their wings so often in their passage, and when ever they settle they also always close their wings. They are found in the months of May and August, as there are two broods a year. The female is seen in the plate at *(h)*. The male is much less.

SMALL SKIPPER.

THE Caterpillar of this Fly is also undiscovered; it flies in woods, and its actions are also similar to the above; but there is only one brood a year, and they appear about the middle of July. The female is figured at *(l)*, which is remarkable for having a black stroke on the middle part of each superior wing, which is wanting in the male.

SCARCE, MARVEL DA JOUR.

THE Caterpillar of this pretty Moth feeds on the oak, and is of a green colour, striped on the side; but not having a drawing of the Caterpillar by me, was obliged to omit it in the Plate. They change to Chrysalis in the earth, and the Moth, which is shewn in the plate at *(e)*, appears in May.

The FRECKLED BROAD BAR.

THE Caterpillar is not known, so that nothing can be said with respect to its colour, but we are certain that it is of the Luper kind, as there was none but Lupers in the cage in which it was bred, all of which was beat from the oak. It lay during the winter in the Chrysalis state, and the Moth was produced in May. It is delineated in the Plate at *(o)*.

The LAPPET.

THE Caterpillar at *(b)*, feeds on black and white thorn, and continues in that state during winter; it is full fed about the letter end of May, when it makes a large spinning, wherein it changes to the Chrysalis seen at *(c)*, and the Moth appears in July. The cock, which is shewn in the Plate at *(a)*, is of a redish chocolate colour, but the hen is much lighter and larger. The best time to look for the Caterpillar is about the beginning of May, on the stems of the food near the ground. They are seldom taken in the Moth state.

EALINGS GLORY.

THE Caterpillar is taken by beating the whitethorn about the middle of May. It changes to the Chrysalis state about the end of the same month, and the Moth appears in September. The Caterpillar is shewn at *(e)*, which is remarkable for a protuberance on the rump. The Chrysalis seen at *(f)*, is of a darkish brown colour. The Moth seen at *(d)*, is seldom taken in that state.

NOVEMBER MOTH.

THE Caterpillars are found by beating, and are very plentiful on the oak in Norwood; they appear about the colour of the bark of the twigs, and are full of sharp pointed protuberances as at *(h)*. They are also sometimes found on the whitethorn, though but seldom. When full fed, which is about the end of September, they spin up in the leaves, and change into Chrysalis, which is represented at *(i)*, and the Moth appears in November; it is shewn in the Plate at *(g)*, this is a male, a figure of the female is needless, as they differ in nothing but the make of their horns, which of the cock are comblike or pectinated, but those of the hen, are like threads.

SNOUT MOTH.

THE Caterpillar, which is figured at *(l)*, is of a pleasant brown or tan colour, and is found by beating oaks in October, when they spin up in the leaves, and change into Chrysalis, and the Moth, which is figured at *(t)*, appears the beginning of June following. They may be taken in plenty, in woody places.

RED NECK.

THIS Moth is commonly seen flying about the tops of high oaks the beginning of June: But the Caterpillar has not yet been discovered. The Moths sometimes fall in the sheet, when beating for Caterpillars. I make no doubt, but, if the oaks were beat about the middle of May, the Caterpillar would be found.

SPRING USHER.

THE Caterpillars feed on oak, and are taken by beating about the middle of May. They are of two sorts, one green spotted with black, the other green, spotted with red. They change to their Chrysalis state in the earth about the twentieth of May, and the Moths appear toward the end of February. The Caterpillars are very plentiful in their season, and are of the Luper kind, as seen at *(m)* and *(n)* in the Plate. The Chrysalis is small and brown as at *(o)*, and the Moth of a lightish brown, mottled with a darker colour as at *(q)*.

Plate XLIII

The name Lappet, Gastropacha quercifolia *may derive from the position of the wings when the moth is at rest. Flat underwings then project beyond forewings, drawn together in a tent-like manner. Another speculation is that the name is due to the unusual flaps or 'lappets' on the sides of the caterpillar. Occurring over the southern half of Britain, the males can be easily attracted by 'assembling'.*

Although found in varying numbers over most of the British Isles, it appears that the Green-brindled Crescent, Allophyes oxyacanthae *was first identified from specimens taken from 'Hedges at Ealing', hence the name by which it was known at that time. One might think Ealing's claim to glory was easily earned by this rather pedestrian insect!*

This is not the familiar November Moth, the slight greyish moth which can be found resting on doors and windows in the damp foggy evenings of late autumn. This is instead a more substantial insect, thought to be the Feathered Thorn, Colotois pennaria, *but Harris's representation of the caterpillar leads one to query whether he confused the species; the caterpillar with the 'sharp pointed protuberances' seems more like that of another 'looper', the Pale Brindled Beauty,* Apocheima pilosaria, *or an allied species.*

The little moth which Harris called the Snout is known to us as the Fan-foot, Herminia tarsipennalis, *a fairly common insect in southern parts of Britain and in Ireland. The caterpillar overwinters when nearly or quite fully grown.*

This is the Red-necked Footman, Atolmis rubricollis. *Harris gave it the name Red Neck, which Haworth changed to the Black Footman, but Harris's name prevailed. Had Harris beaten the oak trees in May he would undoubtedly have collected many caterpillars, but not those of the 'Red Neck'; they feed later in the year, from August to September, on lichens on the branches and trunks of oak and other trees.*

The Spring Usher was called the Early Moth by Haworth, but the name used by Harris has fortunately stayed with us. It is aptly named, as it is one of the earliest to emerge from the chrysalis in the new year and may be found from mid-February to mid-March. The wingless female moth may be seen on tree trunks, usually oak, at that time. Both moths and caterpillars are very variable in colour and markings. This hardy little insect, Agriopis aurantiaria *is found over most of Britain except for northern Scotland.*

Mr Harris del et scul Decemʳ 22 1772

Plate XLIV

This splendid moth, which Harris calls the Spotted Elephant, is known in Britain as a very rare immigrant, the Spurge Hawk-moth, Hyles euphorbiae. He bred it from the chrysalis sent to him from France, perhaps from a friend or correspondent serving abroad, but he was mistaken in thinking the caterpillar he discovered in Kent would have produced a similar moth. This was undoubtedly the fine large caterpillar of the Bedstraw Hawk-moth, Hyles gallii and he was fortunate indeed to obtain it in the marshy ground near Crayford; it was a great mishap that it died. Also an immigrant but rather more frequent, this Hawk-moth has been known temporarily to establish itself for a few years in Britain, the caterpillar feeding on various species of bedstraw and willowherb.

The Grayling, Hipparchia semele is widely distributed, but not everywhere common. The name Tunbridge Grayling, which Harris mentions, was given it by Petiver who said, 'it is very rare about London'. It was first recorded as a British species by Ray. Harris's slight note of disapproval over Wilkes's name for this butterfly echoes the irritation he expressed in the Preface, when Wilkes classified the Burnet moths in 'a new Class of Moths'. The brown-lined, drab mottled caterpillar hibernates when very small, so it was not surprising it was rarely found; but it is not green as he supposed. This fine butterfly prefers rough open ground to fly in, rather than the shrubby woodlands in which Harris found pursuit so difficult, and it occurs in well-scattered locations throughout the British Isles.

The Gatekeeper, Pyronia tithonus is not uncommon over most of England as far north as the Lake District and Ireland. Merrett first described this butterfly, also known today as the Hedge Brown or the Small Meadow Brown, and its pale ochreous-brown caterpillar, unknown to Harris, feeds at night on various grasses.

The Pale-Waved is now known as the Common White Wave, Cabera pusaria. In 1775 Harris gave also the Linnean name lactearia. However, this 'trivial' name should rightly be applied to a little moth with which it shares some resemblance — the Little Emerald, Jodis lactearia. Despite its common name, the Little Emerald retains the delicate green tint for only a short while after emergence; hence the scientific name which is taken from lacteus, meaning milky. C. pusaria can be found in woodland areas in varying numbers, over most of the British Isles. A second brood in July and August increases its numbers in the more favourable climate of southern England.

SPOTTED ELEPHANT.

IT has been long in dispute, whether the spotted Elephant was a native of this island; but it is now past a doubt, as I had the good fortune to find one in marshey ground at Barnscray near Crayford in Kent, about the middle of August. It was better than three inches long, of a dark brown colour; on each ring near the back was a spot of yellowish buff colour, perfectly round, and about the size of a small pea, beneath each of these were two others of the same colour, and about the size of a pin's head; the horn at the tail part, which was about half an inch long, appeared black and glossy. The head was nearly the size of a small pea, and of a lightish yellow brown or tan colour. It was taken up too suddenly at the time it was found, that I could not perceive what it was on, though there appeared nothing near the place, but grass intermixed with clover. Those, however, it would not eat; I tried also with various other herbs to bring it to feed, but my attempts were intirely fruitless, and it died for want. The Chrysalis in the Plate at *(c)* was sent me from Belisle at the time that place was besieged by the English; and the Moth at *(a)* was produced from it about the beginning of June.

GRAILING.

THE Caterpillar it very rarely found, but is well known to feed on grass. It is about an inch and a half long, and of a fine green colour; I should have given a drawing of it in the Plate, but had no draught of it by me at the time I engraved the Plate. They always fly very smartly in woods where there is plenty of shrubs and long grass; and consequently not very easily taken, as the inconveniency of the place hinders pursuit. It is named by Wilks, the Rockunderwing; why he should think it necessary to alter it from its original name, was best known to himself. They were first taken by the Aurelians at Tunbridge, and were called for some time the Tunbridge Grailing. The upper-side of the female is seen at *(d)*, and the under-side at *(e)*, as sitting on the blossom of clover.

GATE KEEPER.

THE haunts of this Fly are on the sides of hedges in lanes and meadows. Their first appearance is about the middle of July, though indeed the females do not appear till the beginning of August; which being taken for a distinct species, went sometime by the name of the Orange Field. Neither the Caterpillar nor Chrysalis has hitherto been discovered. The female is seen in the Plate at *(f)*, shewing the upper-side, the male at *(g)* displaying the under-side, and discovering a great part of the upper side of one of the superior wings, about the middle of which is a dark or black cloud-like spot, which distinguishes the male from the female.

PALE WAVED.

THIS little Moth is taken by beating the hedges in May. The Caterpillar is of the Luper kind, and feeds on whitethorn; it changes to Chrysalis in a spinning about September, and the Moth comes forth about the middle of May. A figure of the upper-side of the Moth is seen in the Plate at *(h)*.

THE END

O God, thou haft taught me from my youth; and hitherto have I declared thy wonderous works.
Pfalm lxxi. ver. 17.